Victor Bryant

METRIC SPACES
Iteration and application

Published by the Press Syndicate of the University of Cambridge
The Pitt Building, Trumpington Street, Cambridge CB2 1RP
40 West 20th Street, New York, NY 10011-4211 USA
10 Stamford Road, Oakleigh, Melbourne 3166, Australia

© Cambridge University Press 1985

First published 1985
Reprinted 1987, 1990, 1994, 1996

Library of Congress catalogue card number: 84-15519

British Library Cataloguing in Publication Data

Bryant, Victor, *1945 –*
Metric spaces: iteration and application
1. Metric spaces
I. Title
515.7'3   QA611.28

ISBN 0 521 31897 1 paperback

Transferred to digital printing 1999

CONTENTS

	Preface	v
1	**Sequences by iteration**	1
1.1	Where are we heading?	1
1.2	Sequences of numbers by iteration	1
1.3	Iterations in a different world	7
2	**Metric spaces**	12
2.1	Distance	12
2.2	Examples of metric spaces	14
2.3	Sequences	22
3	**The three Cs**	29
3.1	Iteration revisited	29
3.2	Closed sets	30
3.3	An internal test for convergence	36
3.4	Complete sets	38
3.5	Compact sets	45
4	**The contraction mapping principle**	52
4.1	Real fixed points	52
4.2	Contractions	57
4.3	Real contractions revisited	62
4.4	Some extensions	64
4.5	Differential equations	71
4.6	The implicit function theorem	83
4.7	Conclusion	86
5	**What makes analysis work?**	87
5.1	Continuity	87
5.2	Attained bounds	92
5.3	Uniform continuity	94
5.4	Inverse functions	96
5.5	Intermediate values	98
5.6	Some final remarks	103
	Index	105

PREFACE

Some years ago I regularly gave a traditional course on metric spaces to second-year special honours mathematics students. I was then asked to give a watered-down version of the same material to a class of combined honours students (who were doing several subjects, including mathematics, at a more general level) but, to put it mildly, the course was not a success. It was impossible to motivate students to generalise real analysis when they had never understood it in the first place and certainly could not remember much of it. It was also counter-productive to start the course by revising real analysis because that convinced the students that this was 'just another analysis course' and their interest was lost for evermore.

So when I gave the course again the following year I decided to turn the material inside out and to *start* with the applications (namely the use of contractions in solving a wide range of equations). This meant that the first chapter was a revision of some iterative techniques used to obtain approximations to solutions of equations. This immediately captured the interest of the class: they enjoyed using their calculators and writing programs to solve the equations. Some of the ideas were entirely new to them; for example using iteration to solve an equation with constraints, or solving a differential equation by iterating with an integral and obtaining a sequence of functions.

The second and third chapters were more traditional but the big difference was that the need for distance, function space, closed set, and so on, had been anticipated and motivated. Another difference was that, having approached the subject via iteration, it was then natural to define all the concepts in terms of sequences: hence closed sets (rather than open ones) formed the basis of the approach.

For most students the fourth chapter was the highlight of the course. It consisted of the contraction mapping principle and the use of its algorithmic proof in solving equations. But I was then able to say to them that, as we had developed all the tools of the subject, it was now an easy matter to look back to real analysis and get a better understanding of it. The last chapter therefore recalled and generalised the classic theorems of real analysis.

This book, then, is an approach to metric spaces along those lines. It tries to avoid assuming that the reader knows much about analysis and if a difficult concept is to be encountered the reader is prepared by several glimpses of it, in examples, well in advance. My aim is to provide a book which can be read and enjoyed by a wide range of second- or third-year students in universities or polytechnics. The only prerequisite is to have done a course on elementary analysis: it is not a prerequisite to have understood it nor to have remembered it all.

There are several people who have contributed indirectly to this work. Firstly I thank Hazel Perfect and John Pym for their most constructive comments. Secondly my well-worn copies of the books on metric spaces by Copson, Simmons and Sutherland testify to the use which they have been to me over the years. Next I must mention my colleagues Mary Hawkins and Harry Burkill: at one time we collaborated on a metric space course, and for any of their examples and proofs which may remain in this new approach I thank them. Also I thank Anne Hall for preparing a beautiful typescript from my almost-illegible original. Finally, and principally, I thank my friend and mentor, Roger Webster. It was his M.Sc. course on functional analysis many years ago which rekindled my pleasure in mathematics and introduced me to metric spaces.

I have tried to provide a readable and natural introduction to an abstract subject in a down-to-earth manner. Even if I can transmit only a fraction of the pleasure which the subject has given me, then I will have been successful.

Sheffield   Victor Bryant
1984

1

Sequences by iteration

1.1 Where are we heading?

Imagine a table top covered with a thin layer of dust. We give it a flick with a duster, but all the dust settles back on the table top. Suppose that after being rearranged every two specks of dust are closer together than they were before. Then the remarkable thing is that there is one and only one speck of dust which is back in the same place that it started in.

That is an example of a set X and a function $f: X \to X$. The property that f brings the points of X closer together is enough to ensure that X has one and only one point which is not moved by f; i.e. a unique $x \in X$ with $x = f(x)$. These so-called 'fixed points' of functions are invaluable in solving equations. For example, the fixed points of the function f given by $f(x) = \frac{1}{3}(x^3 - 5x^2 + 1)$ are those x for which $x = \frac{1}{3}(x^3 - 5x^2 + 1)$; i.e. they are precisely the roots of the equation $x^3 - 5x^2 - 3x + 1 = 0$. In fact, it turns out that this real function f has just one fixed point and that it is easy to find with a calculator; so the unique real root of the cubic equation will be easily found.

In this book our search will be for a large collection of situations X and functions $f: X \to X$ which have unique fixed points. The study of such situations provides an interesting piece of mathematics, but the real fascination will lie in the wide range of problems which they will enable us to solve.

1.2 Sequences of numbers by iteration

To find a real root of the cubic equation
$$x^3 + x^2 - 5x - 3 = 0$$
rewrite the equation as
$$x = \tfrac{1}{5}(x^3 + x^2 - 3)$$
or $x = f(x)$, where $f(x) = \tfrac{1}{5}(x^3 + x^2 - 3)$. Start with a first guess at a

root, 0 for example. Apply f to this first guess to give $f(0) = -0.6$. Apply f to this answer to give $f(-0.6) = -0.5712$. Continue in this way to give a list or sequence of answers:

$$0$$
$$\downarrow$$
$$f(0) = -0.6$$
$$\downarrow$$
$$f(-0.6) = -0.5712$$
$$\downarrow$$
$$f(-0.5712) = -0.5720191$$
$$\downarrow$$
$$f(-0.5720191) = -0.5719924$$
$$\downarrow$$
$$f(-0.5719924) = -0.5719933$$
$$\downarrow$$
$$f(-0.5719933) = -0.5719933$$
$$\downarrow$$
$$\vdots$$

This is the best my calculator can manage, but to this accuracy at least there is no point in continuing as the answer -0.5719933 will keep repeating itself. So (to within my calculator's accuracy)

$$f(-0.5719933) = -0.5719933.$$

In other words, we have found an x with $x = f(x)$ – called a 'fixed point' of f. By our above remarks this x will be (again to within my calculator's accuracy) a root of the cubic equation $x^3 + x^2 - 5x - 3 = 0$. Let us check:

$$(-0.5719933)^3 + (-0.5719933)^2 - 5(-0.5719933) - 3$$
$$= -0.0000000081\ldots,$$

which, of course, is very nearly zero as expected. So having rewritten our equation in the form $x = f(x)$ the solution of the equation was a mechanical process requiring no further thought. But it must be remembered that this process is only finding an approximate root: the sequence which we produce only seems to come to an end because of the limited accuracy of our calculators. And in all cases we ought to substitute our answers back into the equations to check their validity.

Exercise 1 Find a root, to four decimal places, of the cubic equation $x^3 + 2x^2 - 7x - 1 = 0$ by rearranging it into the form $x = f(x) = \frac{1}{7}(x^3 + 2x^2 - 1)$ and making a first guess between -3 and 1. Reapplying f (or 'iterating' with f) should give

Sequences of numbers by iteration

you a sequence settling down (or 'converging') to a root of the equation.

(Calculators are obviously needed! A fairly economical way of calculating $f(x)$, having entered x into the display, is by the following steps:

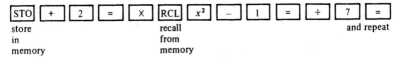

store in memory   recall from memory   and repeat

Of course, a programmable machine can be used to great advantage.)

In Exercise 1 you should have found that one root of the equation $x^3 + 2x^2 - 7x - 1 = 0$ is approximately -0.1378: you could check by substituting this value back into the cubic that it does give a value very near zero.

Now let us examine exactly what was happening in the above examples: we wanted to solve the equation $x = f(x)$ and so we made a first guess x_1 at a root and calculated all future guesses by reapplying f to give

$$x_2 = f(x_1), \; x_3 = f(x_2), \ldots, x_{n+1} = f(x_n), \ldots$$

We then looked at the limit of the sequence

$$x_1, x_2, x_3, \ldots, x_n, \ldots \to x$$

and noted that it had the required property that $x = f(x)$. Of course, as logical people, we must ask some crucial questions about this process:

Will the sequence which we produce by iterating with f definitely converge?

If it does converge will the limit definitely be a root of $x = f(x)$?

If the answers to both these questions were 'yes' and any function led to a sequence which converged to a root of the required equation, then it would be such a wonderful method that practically all other ways of solving equations would be redundant. Let us answer the first question by means of an exercise.

Exercise 2 Repeat the previous exercise by rearranging the cubic equation $x^3 + 2x^2 - 7x - 1 = 0$ into the form $x = f(x) = \frac{1}{7}(x^3 + 2x^2 - 1)$. Try to find a root of the cubic by making a first guess of 2 or more and iterating with f.

Sequences by iteration

Exercise 3 Rearrange $x^3+2x^2-7x-1=0$ into the form $x = f(x) = \sqrt{\{\frac{1}{2}(-x^3+7x+1)\}}$. Try iterating with this f for various first guesses. See if you can find a sequence in this way which converges to a root of the cubic.

([STO] [x^2] [+/−] [+] [7] [=] [×] [RCL] [+] [1] [=] [÷] [2] [=] [$\sqrt{x}$]

and repeat.)

In Exercise 2 you should have got a sequence which very soon became out of hand – clearly not settling down to a root of the cubic. In Exercise 3 if you started with an x_1 between 0 and 2 then you should have got a sequence converging to around 1.91909 – which is very close indeed to a root of the cubic. But for some starting values (such as $x_1 = 3$) you cannot even proceed because $f(3)$ is not defined. So, to summarise, some functions and some starting points will give sequences converging to the required roots, but often a function will lead to a sequence which does not converge. The principles of numerical analysis are beyond our scope here but it is worth mentioning the Newton–Raphson method of iteration, designed as a systematic way of deciding which function to use for iteration: it often leads to a successful iterative process. We content ourselves with one exercise on this.

Exercise 4 To solve the cubic $2x^3+3x^2+6x+1=0$ rearrange it to $6x^3+6x^2+6x = 4x^3+3x^2-1$ and hence $x = f(x) = (4x^3+3x^2-1)/6(x^2+x+1)$. Confirm that

$$f(x) = x - \frac{\text{the original cubic}}{\text{the derivative of the cubic}}$$

(this is how, in general, the f is chosen in the Newton–Raphson method). Choose any x_1, iterate with f and hence find to four decimal places the unique real root of $2x^3+3x^2+6x+1=0$.

The function in Exercise 4 always leads to a convergent sequence, regardless of the first guess x_1. One of the aims of this book is to find a large class of functions which are equally dependable.

Although we have been able to find functions for which the iterative process failed to produce convergent sequences, so far whenever a convergent sequence has been produced it *has* led us to a root. So the answer to our second question

Sequences of numbers by iteration   5

If the sequence produced by iteration with f converges will the limit definitely be a root of $x = f(x)$?

seems much more likely to have a favourable answer. Imagine, for example, that we are trying to find an x for which $x = f(x)$, where f is a cubic in x, and we produce a convergent sequence.

$$x_1, x_2 = f(x_1), x_3 = f(x_2), \ldots, x_{n+1} = f(x_n), \ldots \to x.$$

Then $f(x_n)$ is simply a cubic in x_n and so as the x_ns get closer to x so do the $f(x_n)$s get closer to $f(x)$; i.e.

$$
\begin{array}{cccc}
f(x_1), & f(x_2), & f(x_3), \ldots, & f(x_n), \ldots \to f(x) \\
\| & \| & \| & \| \\
x_2, & x_3, & x_4, \ldots, & x_{n+1}, \ldots \, (\to x\,!!!)
\end{array}
$$

But, as you can see, the sequence of $f(x_n)$s is simply part of the sequence of x_ns, which converge to x. Therefore $x = f(x)$ and the limit x does satisfy the equation.

Most of the functions which we encounter will be 'continuous'; i.e. if x and y are sufficiently close, then $f(x)$ and $f(y)$ are close. If f is of this type and an iterative process leads to a convergent sequence, then the limit of that sequence must be a root of $x = f(x)$, as we now see.

Theorem 1.1 Let X be a subset of the real numbers $\mathbb{R}$, let $f: X \to X$ be continuous and let $x_1 \in X$. Then if the sequence

$$x_1, x_2 = f(x_1), x_3 = f(x_2), \ldots, x_{n+1} = f(x_n), \ldots$$

converges to $x \in X$ it follows that $x = f(x)$.

Proof Let $x_2 = f(x_1)$, $x_3 = f(x_2)$, etc. and assume that

$$x_1, x_2, x_3, \ldots, x_n, \ldots \to x$$

as stated. Then the x_ns get very close to x and so (by continuity) the $f(x_n)$s get very close to $f(x)$; i.e.

$$
\left.
\begin{array}{cccc}
f(x_1), & f(x_2), & f(x_3), \ldots, & f(x_n), \ldots \to f(x) \\
\| & \| & \| & \| \\
x_2, & x_3, & x_4, & x_{n+1}, \ldots \to x
\end{array}
\right\} \therefore x = f(x). \quad \square
$$

The above result is only a very special case of one which we shall meet later (where we will also discuss the necessary conditions on the domain X of f).

For those interested enough to check that the condition of continuity of f is actually needed in the above theorem we will consider a slightly more unusual example (the reader not interested in this esoteric point can turn immediately to the goat example below – or indeed to the beginning of Section 1.3 if he has had enough of solving equations by producing real sequences). Let $[x]$ denote the 'largest integer not exceeding x'. For example $[3.142] = 3$. This is one of the simpler functions which is not continuous, for it is possible for x to be very close to y (for example 1.99 and 2) without $[x]$ being close to $[y]$. Now let us try to solve the equation

$$x = f(x) = [x] + 1 - \tfrac{1}{2}([x] + 1 - x)^2$$

by an iterative process starting with $x_1 = 1$. This gives

$x_2 = f(x_1) = [1] + 1 - \tfrac{1}{2}([1] + 1 - 1)^2 = 2 - \tfrac{1}{2} = 1.5,$
$x_3 = f(x_2) = [1.5] + 1 - \tfrac{1}{2}([1.5] + 1 - 1.5)^2 = 2 - \tfrac{1}{2}(0.5)^2 = 1.875$ etc.

which leads to the sequence

1, 1.5, 1.875, 1.9921875, 1.9999695, ...

which is clearly converging to 2. So

$f(1), f(1.5), f(1.875), f(1.9921875), \ldots$
∥ ∥ ∥ ∥
1.5, 1.875, 1.9921875, 1.9999695, ...

also converges to 2. But

$f(1), f(1.5), f(1.875), \ldots \nrightarrow f(2)$

since $f(2) = 2.5$. So the limit does not satisfy $x = f(x)$ and the continuity of f is needed in the above theorem.

To conclude our series of numerical examples we will solve a very famous problem concerning a goat: the complicated function with which we will iterate begins to show the power of this method. The problem is that a farmer owns a circular plot of land (of unit diameter, say) which is covered in grass. He wants to tie his goat by a long rope, one end attached to a point on the circumference of the field and the other end attached to the goat. How long should the rope be in order that the goat can eat exactly half the grass?

Exercise 5 Show (or take my word for it) that the shaded area in Figure 1.1 opposite is

$$\tfrac{1}{2}\sin^{-1} x - \tfrac{x}{2}(1 - x^2)^{1/2} + x^2 \cos^{-1} x.$$

Hence find an equation in x if the shaded area is equal to half of the area of the larger circle. By rearranging your equation into the form $x = f(x)$ for some f and by choosing a suitable

Iterations in a different world

first guess for x_1, obtain a sequence which converges to the required value of x.

Note that the equation you get in Exercise 5 may have other roots besides the one relevant to the goat problem, so make a quick mental check that the limit of your sequence is the right sort of size. The rearrangement which I used was obtained by finding an expression for $\sin^{-1} x$ and hence for x: the iterative process with a starting value of just over $\frac{1}{2}$ then produced a convergent sequence (with limit of approximately 0.57935).

1.3 Iterations in a different world

The general principle behind our iterative techniques in the previous section was

old guess ⟶ [process] ⟶ new guess

where the process can be repeated to give a sequence of improving guesses. In all the examples so far the process has been dealing with numbers, but now let us illustrate a process which deals with something different.

Note before proceeding that the list of numbers

  3, 2, 1, 1, 0, 0, 0

is an inventory of itself, in the sense that

3	2	1	1	0	0	0
‖	‖	‖	‖	‖	‖	‖
no. of 0s in the list	no. of 1s in the list	no. of 2s in the list	no. of 3s in the list	no. of 4s in the list	no. of 5s in the list	no. of 6s in the list.

We are going to find a self-counting list of ten numbers.

Fig. 1.1

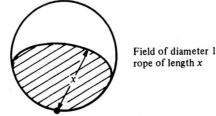

Field of diameter 1
rope of length x

Sequences by iteration

So let $\mathbb{R}^{10}$ be the set of points with ten real coordinates (or 'lists of ten real numbers'). We shall find a member $\mathbf{x}$ of $\mathbb{R}^{10}$ with

$\mathbf{x} = (x_0, \quad x_1, \quad x_2, \quad x_3, \quad x_4, \quad x_5, \quad x_6, \quad x_7, \quad x_8, \quad x_9)$
$\quad\quad\ \ \parallel \quad\quad \parallel$
no. of   no. of   etc.
0s in   1s in
the   the
list   list

In order to employ an iterative technique we will use the process whereby given any member of $\mathbb{R}^{10}$ we count up the occurrences of the numbers 0 to 9 as its coordinates. For example,

$(8, 7, 8, 2, 1, 1\frac{1}{3}, 5.9, 0, 1, 1) \xrightarrow{f\ \text{count up occurrences of 0 to 9}} (1, 3, 1, 0, 0, 0, 0, 1, 2, 0).$

Repeatedly applying the process f to give a sequence of points of $\mathbb{R}^{10}$ yields, for the above starting point,

$\mathbf{x}_1 = (8, 7, 8, 2, 1, 1\frac{1}{3}, 5.9, 0, 1, 1),$
$\mathbf{x}_2 = f(\mathbf{x}_1) = (1, 3, 1, 0, 0, 0, 0, 1, 2, 0),$
$\mathbf{x}_3 = f(\mathbf{x}_2) = (5, 3, 1, 1, 0, 0, 0, 0, 0, 0),$
$\mathbf{x}_4 = f(\mathbf{x}_3) = (6, 2, 0, 1, 0, 1, 0, 0, 0, 0),$
$\mathbf{x}_5 = f(\mathbf{x}_4) = (6, 2, 1, 0, 0, 0, 1, 0, 0, 0),$
$\mathbf{x}_6 = f(\mathbf{x}_5) = (6, 2, 1, 0, 0, 0, 1, 0, 0, 0).$

So already the process f is repeating itself and we have actually reached a point

$\mathbf{x} = (6, 2, 1, 0, 0, 0, 1, 0, 0, 0)$

for which $f(\mathbf{x}) = \mathbf{x}$; i.e. $\mathbf{x}$ is a self-counting list. Try this process for yourself with different starting points (and with lists of different lengths).

The above example illustrated that our iterations need not only deal with single numbers. Our next few examples will concern iteration with functions and this will have far-reaching consequences. To start with, imagine that we have a process which takes in some function x (e.g. $x(t) = t^2$ for $t \in \mathbb{R}$) and, after processing, gives out a new function y (e.g. $y(t) = t^2 + 1$ for $t \in \mathbb{R}$). In general

Iterations in a different world

$$x \longrightarrow \boxed{\text{process}} \longrightarrow \text{a new function } y.$$

(a function $\mathbb{R} \to \mathbb{R}$)

Since y is a function, to define it we must stipulate what $y(t)$ is in terms of x's values which we know all about. For example, our process could start with x and produce y with, for each $t \in \mathbb{R}$,

$$y(t) = 1 + \int_0^t u^2 x(u) \, du.$$

So if x is the zero function, then $y(t) = 1$ for all t; and if $x(t) = \sin(t^3)$, then

$$y(t) = 1 + \int_0^t u^2 \sin(u^3) \, du = 1 + [-\tfrac{1}{3} \cos(u^3)]_0^t$$
$$= \tfrac{4}{3} - \tfrac{1}{3} \cos(t^3).$$

This process can be applied to any continuous function (so that the integral is defined) to produce another continuous function. So if X is the set of continuous functions (from $\mathbb{R}$ to $\mathbb{R}$, say) then the process takes an $x \in X$ and creates a $y \in X$: the process is simply a function $f: X \to X$. This is no different in principle from the examples in the previous section except that the elements of X are themselves functions. To summarise, the above process is a function $f: X \to X$ where $x \in X$ is taken to $y = f(x)$ given by

$$(f(x))(t) = y(t) = 1 + \int_0^t u^2 x(u) \, du.$$

Exercise 6 Let x be the function given by $x(t) = 1$. Evaluate the function $f(x)$; i.e. find $(f(x))(t)$ in terms of t, where f is as stated above.

In Exercise 6 $f(x)$ turned out to be the function given by $(f(x))(t) = 1 + \tfrac{1}{3} t^3$.

Exercise 7 Let x_1 be the function given by $x_1(t) = 1$. Let $f(x_1) = x_2$ where f is as above. Then x_2 is the function given by $x_2(t) = 1 + \tfrac{1}{3} t^3$. Let $f(x_2) = x_3$: evaluate $x_3(t)$. Let $f(x_3) = x_4$: evaluate $x_4(t)$.

We learnt in the previous section of the possible interest of reapplying a function and investigating the behaviour of the resulting sequence. If we consider our new f defined as above and start with the function

Sequences by iteration

$x_1(t) = 1$, then reapplying f gives the sequence

$$x_1(t) = 1, \quad x_2(t) = 1 + \frac{t^3}{3}, \quad x_3(t) = 1 + \frac{t^3}{3} + \frac{t^6}{9 \times 2!},$$

$$x_4(t) = 1 + \frac{t^3}{3} + \frac{t^6}{9 \times 2!} + \frac{t^9}{27 \times 3!}, \ldots$$

As yet, we have no formal way of testing whether a sequence of functions settles down, but you may have already spotted a pattern in the above sequence and guessed that it is settling down to the function given by

$$x(t) = 1 + \frac{t^3}{3} + \frac{t^6}{3^2 \times 2!} + \frac{t^9}{3^3 \times n!} + \cdots + \frac{t^{3n}}{3^n \times n!} + \cdots = e^{t^3/3}.$$

Our work in the previous section might lead us to suspect that this function is a fixed point of f; i.e. $f(x) = x$ or

$$x(t) = (f(x))(t) = 1 + \int_0^t u^2 x(u) \, du \quad (\text{all } t \in \mathbb{R}).$$

And indeed this is the case, as can be checked by integration.

So it seems that we have solved the equation in x,

$$x(t) = 1 + \int_0^t u^2 x(u) \, du \quad (t \in \mathbb{R}),$$

by starting with any function x_1 and reapplying f to give a sequence of functions: their limit was the required root.

At the moment much of this is informal: we do not really know what it means to say that a sequence of functions converges. (Nor have we seen the significance of equations like the integral equation above.) It seems that the general principle of iteration with a function f might be useful in many situations apart from solving numerical equations. But we must develop the theory for coping with functions $f \colon X \to X$ and for determining in the most general situations whether the sequence

$$x_1, \quad x_2 = f(x_1), \quad x_3 = f(x_2), \quad \ldots$$

converges. Before doing so, we conclude this section with one further 'functional' example of the above type.

Exercise 8 Let X be the set of continuous real functions defined on the open interval $]-1, 1[$. (We shall always use the notation $]a, b[$ for the open interval $\{x \in \mathbb{R} : a < x < b\}$.) For $x \in X$ let $f(x) \in X$ be defined by

$$(f(x))(t) = 1 - \int_0^t [x(u)]^2 \, du.$$

Let x_1 be the zero function (i.e. $x_1(t) = 0$ for all $t \in\,]-1, 1[$) and let $x_2 = f(x_1)$, $x_3 = f(x_2)$, etc. Show that $x_2(t) = 1$ and $x_3(t) = 1 - t$. Evaluate x_4 and x_5 and note that the functions $x_1, x_2, x_3, \ldots$ seem to be approaching the function $x \in X$ given by

$$x(t) = 1 - t + t^2 - t^3 + \cdots = \frac{1}{1+t}.$$

Confirm by direct integration that $x(t) = 1/(1+t)$ is indeed a root of the equation

$$x(t) = 1 - \int_0^t [x(u)]^2 \, du.$$

2

Metric spaces

2.1 Distance

Our eventual aim is to be able to consider sequences (sometimes of numbers, sometimes of functions and sometimes, perhaps, from other worlds) and to ask whether they settle down or converge. To ask whether the members of X

$$x_1, x_2, x_3, \ldots, x_n, \ldots$$

get closer to x we need to be able to say how far x_n is from x. Hence our basic need is for the idea of the distance between any two members of X.

The approach we choose is a pure mathematical one, but it happens to pay off handsomely in some of the later applications. We shall isolate the three fundamental properties of distance and base all our deductions on these three properties alone. This makes proofs much easier (with only three properties to think about rather than all the technical properties of $\mathbb{R}$, say) and it makes the results much more general because they apply to any situation which has a concept of distance satisfying our three basic properties.

The three properties are apparent from a normal mileage chart (Figure 2.1). The first obvious property is that zeros appear down the diagonal and nowhere else; so the distance between x and y is zero if and only if $x = y$. The second obvious property of the mileage chart is its symmetry; i.e. the distance from x to y is the same as that from y to x. The third property is inherent in the chart but is not so immediately seen. The distance from London to Manchester is 190 miles, whereas

Fig. 2.1

	London	Manchester	Sheffield
London	0	190	160
Manchester	190	0	40
Sheffield	160	40	0

Distance

the distance from London to Sheffield to Manchester is $160+40=200$ miles. In general, if we go via some other point we must travel as far or further than we would on the direct route (Figure 2.2). With our usual straight-line distances this property can be illustrated by means of the three vertices of a triangle: for this reason it is known as the *triangle inequality*.

Fig. 2.2

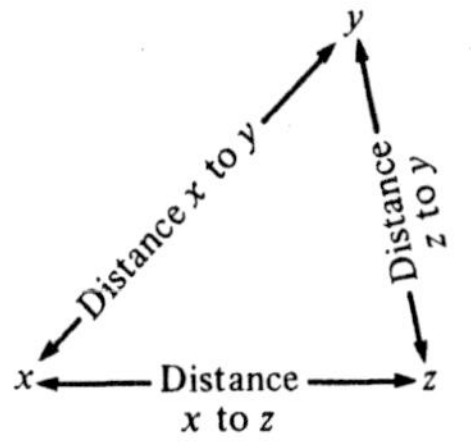

distance x to y
$\leqslant$ (distance x to z) + (distance z to y)

Definition Let X be a non-empty set and for each $x, y \in X$ let $d(x, y)$ be a real number satisfying

M1. $d(x, y) = 0$ if and only if $x = y$;
M2. $d(x, y) = d(y, x)$ for each $x, y \in X$;
M3. $d(x, y) \leqslant d(x, z) + d(z, y)$ for each x, y and $z \in X$.

Then d is called a *metric* or *distance* on X, and X together with d is called a *metric space* (X, d).

Of course, there are other properties of d which you might have expected to be in the definition. For example, $d(x, y) \geqslant 0$ for all x and y. But this property follows from the other three, as the interested reader might like to confirm.

Exercise 9 Use the fact that
$$d(x, x) \leqslant d(x, y) + d(y, x)$$
for any x, y in a metric space (X, d) to deduce that $d(x, y) \geqslant 0$.

Note also that if distances are generally greater going via an additional point, then they are greater going via any number of additional points $z_1, z_2, \ldots, z_n$: for, by repeated use of M3,

$$\begin{aligned}
d(x, y) &\leqslant d(x, z_1) + d(z_1, y) \\
&\leqslant d(x, z_1) + d(z_1, z_2) + d(z_2, y) \\
&\leqslant d(x, z_1) + d(z_1, z_2) + d(z_2, z_3) + d(z_3, y) \\
&\leqslant \cdots \\
&\leqslant d(x, z_1) + d(z_1, z_2) + d(z_2, z_3) + \cdots + d(z_n, y).
\end{aligned}$$

14   Metric spaces

It is crucial to our later applications that there are many examples of metric spaces. To confirm that some of the examples satisfy $M3$ may be a little technical, but that does not alter the fact that the basic idea of distance and a metric space is a very simple and elegant one.

2.2   Examples of metric spaces

1. Let $X = \mathbb{R}$, the set of real numbers, and for $x, y \in X$ define $d(x, y)$ by $d(x, y) = |x - y|$. Then (X, d) is a metric space, as we now verify:

$M1$. $d(x, y) = |x - y| = 0$ if and only if $x = y$;
$M2$. $d(x, y) = |x - y| = |-(x - y)| = |y - x| = d(y, x)$;
$M3$. $d(x, y) = |x - y| = |(x - z) + (z - y)|$
$\leqslant |x - z| + |z - y|$
$= d(x, z) + d(z, y)$.

The verification of $M3$ in Example 1 used the easy fact that, for real numbers, $|a + b| \leqslant |a| + |b|$. This same property is true for complex numbers a and b: in fact (as the interested reader can verify for himself), if the arguments of a, b and $a + b$ are θ, ϕ and ψ respectively, then

$$|a + b| = |a| \cos(\theta - \psi) + |b| \cos(\phi - \psi) \leqslant |a| + |b|.$$

So precisely the same reasoning which showed that $d(x, y) = |x - y|$ defines a metric for $\mathbb{R}$ will show that it defines a metric for $\mathbb{C}$. Hence:

2. Let $X = \mathbb{C}$, the set of complex numbers, and for $x, y \in X$ define $d(x, y)$ by $d(x, y) = |x - y|$. Then, as above, (X, d) is a metric space.

3. Let $X = \mathbb{R}^2$, the set of points in the coordinate plane. We shall use bold print for the members $\mathbf{x}$ of X to distinguish them from their coordinates; $\mathbf{x} = (x_1, x_2)$, etc. For $\mathbf{x}, \mathbf{y} \in X$ define $d(\mathbf{x}, \mathbf{y})$ by

$$d(\mathbf{x}, \mathbf{y}) = d((x_1, x_2), (y_1, y_2)) = [(x_1 - y_1)^2 + (x_2 - y_2)^2]^{1/2}.$$

This is precisely the usual straight-line or Euclidean distance between points in the plane. It is intuitively clear that our everyday distance satisfies $M1$, $M2$ and $M3$, but we now give the formal verification:

$M1$. $d(\mathbf{x}, \mathbf{y}) = 0$ if and only if $[(x_1 - y_1)^2 + (x_2 - y_2)^2]^{1/2} = 0$ which happens if and only if $x_1 = y_1$ and $x_2 = y_2$, i.e. $\mathbf{x} = \mathbf{y}$.

Examples of metric spaces

M2. $d(\mathbf{x},\mathbf{y}) = [(x_1-y_1)^2 + (x_2-y_2)^2]^{1/2}$
$= [(y_1-x_1)^2 + (y_2-x_2)^2]^{1/2} = d(\mathbf{y},\mathbf{x})$.

M3. We will establish M3 for $\mathbf{x} = (x_1, x_2)$, $\mathbf{y} = (y_1, y_2)$ and $\mathbf{z} = (z_1, z_2)$.

Note that the quadratic in t given by
$$[(x_1-z_1)^2 + (x_2-z_2)^2]t^2 + 2[(x_1-z_1)(z_1-y_1) + (x_2-z_2)(z_2-y_2)]t + [(z_1-y_1)^2 + (z_2-y_2)^2]$$
is in fact
$$[(x_1-z_1)t + (z_1-y_1)]^2 + [(x_2-z_2)t + (z_2-y_2)]^2$$
which is never negative for real t. But if a quadratic $\alpha t^2 + \beta t + \gamma$ with $\alpha, \gamma \geq 0$ is never negative for real t then its discriminant is not positive, and $\beta \leq 2(\alpha\gamma)^{1/2}$. So in this case
$$2[(x_1-z_1)(z_1-y_1) + (x_2-z_2)(z_2-y_2)]$$
$$\leq 2\{[(x_1-z_1)^2 + (x_2-z_2)^2][(z_1-y_1)^2 + (z_2-y_2)^2]\}^{1/2}$$
$$= 2d(\mathbf{x},\mathbf{z})d(\mathbf{z},\mathbf{y}).$$

Therefore
$$[d(x,y)]^2 = (x_1-y_1)^2 + (x_2-y_2)^2$$
$$= [(x_1-z_1) + (z_1-y_1)]^2 + [(x_2-z_2) + (z_2-y_2)]^2$$
$$= (x_1-z_1)^2 + (x_2-z_2)^2 + 2[(x_1-z_1)(z_1-y_1)$$
$$+ (x_2-z_2)(z_2-y_2)] + (z_1-y_1)^2 + (z_2-y_2)^2$$
$$\leq (x_1-z_1)^2 + (x_2-z_2)^2 + 2d(\mathbf{x},\mathbf{z})d(\mathbf{z},\mathbf{y}) + (z_1-y_1)^2 + (z_2-y_2)^2$$
$$= [d(\mathbf{x},\mathbf{z})]^2 + 2d(\mathbf{x},\mathbf{z})d(\mathbf{z},\mathbf{y}) + [d(\mathbf{z},\mathbf{y})]^2$$
$$= [d(\mathbf{x},\mathbf{z}) + d(\mathbf{z},\mathbf{y})]^2$$
and
$$d(\mathbf{x},\mathbf{y}) \leq d(\mathbf{x},\mathbf{z}) + d(\mathbf{z},\mathbf{y})$$
as required. So (X, d) is a metric space.

Readers familiar with the representation of $\mathbb{C}$ in an Argand diagram (or 'complex plane') will perhaps suspect that there is a strong connection between distances in $\mathbb{C}$ and distances in $\mathbb{R}^2$. In fact if $x = x_1 + x_2 i$ and $y = y_1 + y_2 i$ in $\mathbb{C}$ (where x_1, x_2, y_1 and y_2 are real), then the definition of distance d in $\mathbb{C}$ given in Example 2 shows that
$$d(x,y) = |(x_1 + x_2 i) - (y_1 + y_2 i)|$$
$$= |(x_1 - y_1) + (x_2 - y_2)i|$$
$$= [(x_1 - y_1)^2 + (x_2 - y_2)^2]^{1/2},$$
which is precisely the distance from (x_1, x_2) to (y_1, y_2) in $\mathbb{R}^2$ using the

definition of distance from Example 3. So the verification of $M3$ in that example also provides a verification for $\mathbb{C}$ (or vice versa).

4. Let $X = \mathbb{R}^n$, the set of points with n real coordinates; e.g. $\mathbf{x} = (x_1, x_2, \ldots, x_n) \in X$. (In the case $n = 2$ this gives the coordinate plane of Example 3 and in the case $n = 3$ this gives the usual 3-dimensional space.) For $\mathbf{x}, \mathbf{y} \in X$ define $d(\mathbf{x}, \mathbf{y})$ by

$$d(\mathbf{x}, \mathbf{y}) = d((x_1, x_2, \ldots, x_n), (y_1, y_2, \ldots, y_n))$$
$$= [(x_1 - y_1)^2 + (x_2 - y_2)^2 + \cdots + (x_n - y_n)^2]^{1/2}.$$

Then (X, d) is a metric space. The verification is very similar to the case $n = 2$ given in Example 3, except, of course, that $M3$ is verified by considering the quadratic

$$[(x_1 - z_1)^2 + \cdots + (x_n - z_n)^2]t^2 + 2[(x_1 - z_1)(z_1 - y_1) + \cdots + (x_n - z_n)(z_n - y_n)]t + [(z_1 - y_1)^2 + \cdots + (z_n - y_n)^2].$$

Exercise 10 Verify that the example (X, d) in **4** does satisfy $M1$, $M2$ and $M3$ and hence that it is a metric space.

The examples so far have simply put into the setting of a metric space all the familiar concepts of distance (between real numbers or points in a plane, etc.) but the main advantage of our abstraction will be that it incorporates some far less familiar situations, some of which will lead to powerful applications. The first of these unusual examples shows how our metrics can differ considerably from an intuitive concept of distance.

5. Let X be any non-empty set and for $x, y \in X$ define $d(x, y)$ by

$$d(x, y) = \begin{cases} 0 & \text{if } x = y \\ 1 & \text{if } x \neq y. \end{cases}$$

Then (X, d) is a metric space, called a *discrete space*.

Before we leave the reader to verify that a discrete space is a metric space some comments about $M3$ might be helpful. If X is a set and $d(x, y)$ is defined for each $x, y \in X$ such that $d(x, y) \geq 0$ and d satisfies $M1$ and $M2$, then it is easy to check that

$$d(x, x) \leq d(x, z) + d(z, x) \quad \text{(which is } M3 \text{ with } x = y\text{)}$$
$$d(x, y) \leq d(x, y) + d(y, y) \quad \text{(which is } M3 \text{ with } y = z\text{)}$$
$$d(x, y) \leq d(x, x) + d(x, y) \quad \text{(which is } M3 \text{ with } x = z\text{)}.$$

In other words, if $d(x, y) \geq 0$ and $M1$, $M2$ hold, then $M3$ holds if any

Examples of metric spaces

two out of x, y and z are the same. So it is only necessary to check $M3$ for different x, y and z.

Exercise 11 Verify that a discrete space is a metric space.

Given the points $(0, 2)$ and $(3, 6)$ in $\mathbb{R}^2$ how far apart are they? In the usual straight-line measurement of distance (as in Example 3) they are 5 apart, but in the discrete measurement of distance (as in Example 5) they are 1 apart. Since sets can have more than one metric on them we ought to make it clear which metric we are referring to when asking any question concerning distance. However, unless we say anything to the contrary, when talking about distances in $\mathbb{R}$, $\mathbb{R}^2$, $\mathbb{R}^n$ or $\mathbb{C}$ we shall assume that the usual metrics outlined in the above examples are being used.

By way of a diversion we introduce yet another metric on $\mathbb{R}^2$.

6. Let $X = \mathbb{R}^2$ and for $\mathbf{x}, \mathbf{y} \in X$ define $d(\mathbf{x}, \mathbf{y})$ by

$$d(\mathbf{x}, \mathbf{y}) = d((x_1, x_2), (y_1, y_2)) = \begin{cases} |x_1 - y_1| & \text{if } x_2 = y_2 \\ |x_1| + |x_2 - y_2| + |y_1| & \text{if } x_2 \neq y_2. \end{cases}$$

Then (X, d) is a metric space. We verify $M1$, $M2$ and $M3$:

$M1$. $d(\mathbf{x}, \mathbf{y}) = 0$ if and only if $|x_1 - y_1| = 0$ and $x_2 = y_2$ or $|x_1| + |x_2 - y_2| + |y_1| = 0$ and $x_2 \neq y_2$. These latter conditions are incompatible and so $d(\mathbf{x}, \mathbf{y}) = 0$ if and only if $x_1 = y_1$ and $x_2 = y_2$; i.e. $\mathbf{x} = \mathbf{y}$.

$M2$.

$$d(\mathbf{x}, \mathbf{y}) = \begin{cases} |x_1 - y_1| & \text{if } x_2 = y_2 \\ |x_1| + |x_2 - y_2| + |y_1| & \text{if } x_2 \neq y_2 \end{cases}$$

$$= \begin{cases} |y_1 - x_1| & \text{if } y_2 = x_2 \\ |y_1| + |y_2 - x_2| + |x_1| & \text{if } y_2 \neq x_2 \end{cases} = d(\mathbf{y}, \mathbf{x}).$$

$M3$. To verify $M3$ for $\mathbf{x} = (x_1, x_2)$, $\mathbf{y} = (y_1, y_2)$ and $\mathbf{z} = (z_1, z_2)$ we note firstly that $|x_1 - y_1| \leq d(\mathbf{x}, \mathbf{y})$ and then check $M3$ in cases.

If $x_2 = y_2$, then

$d(\mathbf{x}, \mathbf{y}) = |x_1 - y_1|$
$\leq |x_1 - z_1| + |z_1 - y_1|$
$\leq d(\mathbf{x}, \mathbf{z}) + d(\mathbf{z}, \mathbf{y}).$

If $x_2 \neq y_2$, then z_2 cannot equal both x_2 and y_2, so assume that $z_2 \neq x_2$ (the cases $z_2 \neq y_2$ following similarly). Then

17

18   *Metric spaces*

$$d(\mathbf{x}, \mathbf{y}) = |x_1| + |x_2 - y_2| + |y_1|$$
$$\leq |x_1| + |x_2 - z_2| + |z_2 - y_2| + |y_1|$$
$$\leq \begin{cases} (|x_1| + |x_2 - z_2| + |z_1|) + |z_1 - y_1| & \text{if } y_2 = z_2 \\ (|x_1| + |x_2 - z_2| + |z_1|) + (|z_1| + |z_2 - y_2| + |y_1|) & \text{if } y_2 \neq z_2 \end{cases}$$
$$= d(\mathbf{x}, \mathbf{z}) + d(\mathbf{z}, \mathbf{y}),$$
and M3 is verified.

So this is yet another metric on $\mathbb{R}^2$. Its distances can be illustrated in the plane: we show both cases in one illustration (Figure 2.3). Perhaps the picture shows why this metric is known as the 'lift metric' or the 'raspberry pickers' metric'! (Think about travelling between parts of a tall building or between parts of a field planted with rows of raspberry plants but with a central path linking the rows.)

Fig. 2.3

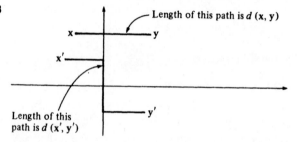

Exercise 12 Let $X = \mathbb{C}$, the set of complex numbers, and define d by
$$d(x, y) = \begin{cases} 0 & \text{if } x = y \\ |x| + |y| & \text{if } x \neq y. \end{cases}$$
Show that d is a metric for $\mathbb{C}$ and think of a suitably illuminating name for it.

Exercise 13 Let $X = \mathbb{R}^2$ and define d by
$$d(\mathbf{x}, \mathbf{y}) = d((x_1, x_2), (y_1, y_2)) = \max\{|x_1 - y_1|, |x_2 - y_2|\};$$
i.e. the larger of $|x_1 - y_1|$ or $|x_2 - y_2|$. Show that d is a metric for $\mathbb{R}^2$. (The keen reader can check that
$$d(\mathbf{x}, \mathbf{y}) = d((x_1, \ldots, x_n), (y_1, \ldots, y_n))$$
$$= \max\{|x_1 - y_1|, \ldots, |x_n - y_n|\}$$
defines a metric for $\mathbb{R}^n$.)

Examples of metric spaces

Those last few non-standard examples show that the abstract concept of distance can differ from our preconceived ideas of distance. But by far the most important examples for our purposes concern the distance between functions.

Consider the functions $x, y: [0, 2] \to \mathbb{R}$ given by $x(t) = t^2$ and $y(t) = t+1$. How far apart are these functions? How could we sensibly measure this distance? A good starting point is with a picture, so let us look at the graphs of these functions (Figure 2.4). One way of

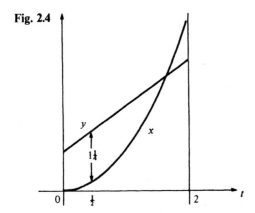

Fig. 2.4

measuring their distance apart is to ask how far their graphs are ever apart. In this case when $t = \frac{1}{2}$ the graphs are $1\frac{1}{4}$ apart, and that is the maximum separation which occurs. In general, given two functions x and y with domain A we ought to consider the values of $|x(t) - y(t)|$ as t takes all values in A and then ask what the biggest of these values is. Since we shall only ever be considering the distance between continuous functions on closed intervals $[a, b]$ ($= \{x \in \mathbb{R}: a \leqslant x \leqslant b\}$) this idea of the 'biggest separation' turns out to be sufficient for our needs: the reader who wishes to consider why this idea is not sufficient for all functions can read the comments after the next example.

7. Let X be the set of continuous functions from $[a, b]$ to $\mathbb{R}$, and for $x, y \in X$ define $d(x, y)$ by

$$d(x, y) = \max \{|x(t) - y(t)|: t \in [a, b]\}.$$

Then (X, d) is a metric space, as we now verify:

M1. $d(x, y) = \max \{|x(t) - y(t)|: t \in [a, b]\} = 0$ if and only if $|x(t) - y(t)| = 0$ for each $t \in [a, b]$; i.e. if and only if $x(t) = y(t)$ for each $t \in [a, b]$; i.e. if $x = y$.

M2. $d(x, y) = \max \{|x(t) - y(t)|: t \in [a, b]\}$
$= \max \{|y(t) - x(t)|: t \in [a, b]\}$
$= d(y, x).$

M3. $d(x, y) = \max \{|x(t) - y(t)|: t \in [a, b]\}$
$= |x(t_0) - y(t_0)|$ for some $t_0 \in [a, b]$
$\leqslant |x(t_0) - z(t_0)| + |z(t_0) - y(t_0)|$
$\leqslant \max \{|x(t) - z(t)|: t \in [a, b]\}$
$+ \max \{|z(t) - y(t)|: t \in [a, b]\}$
$= d(x, z) + d(z, y),$

as required.

Exercise 14 Let $x, y: [0, 1] \to \mathbb{R}$ be given by $x(t) = t$ and $y(t) = t^2$. Calculate the distance between x and y in the metric space of Example 7.

In general, we shall let $C(a, b)$ denote the collection of continuous functions from $[a, b]$ to $\mathbb{R}$. (Those readers who find it hard to consider a set of functions can think of it as a set of graphs.) Unless we say otherwise, the metric which we shall be using on this set is the 'max' metric defined in 7. Readers who wish to pursue this subject further will find this metric referred to elsewhere as the 'sup' metric. We shall pause now to explain some of the possible inadequacies of the 'max' metric approach in more general situations. The reader not interested in this subtle point can turn immediately to Example 9.

What is the distance between the functions x, y from the open interval $]0, 1[$ to $\mathbb{R}$ given by $x(t) = t$ and $y(t) = t^2 + t + 1$? The above approach would imply that we should look at

$$|x(t) - y(t)| = |t - (t^2 + t + 1)| = t^2 + 1$$

for $t \in]0, 1[$ and ask what is its biggest value. But $t^2 + 1$ ($t \in]0, 1[$) takes all values between 1 and 2 without actually ever equalling 2. So instead of asking for the maximum of $|x(t) - y(t)|$ (which never actually occurs) we must ask for the smallest number which is at least as big as every $|x(t) - y(t)|$. In general a non-empty set E of real numbers is *bounded above* if there exists a number u (an *upper bound*)

Examples of metric spaces

with $e \leqslant u$ for all $e \in E$. Then E has a **least** upper bound, called its *supremum* or sup E. If E has a biggest member then that biggest member is the supremum.

Similarly a set E is *bounded below* if there exists a number l (a *lower bound*) with $l \leqslant e$ for all $e \in E$, and E is *bounded* if it is bounded above and below.

Now we can extend Example 7 to include functions $x: A \to \mathbb{R}$ provided that $\{x(t): t \in A\}$ is bounded: this simply ensures that the distance between two functions is never infinite.

8. Let X be the set of bounded functions from some set A to $\mathbb{R}$; i.e. those $x: A \to \mathbb{R}$ such that the set $\{x(t): t \in A\}$ is bounded. For $x, y \in X$ define $d(x, y)$ by
$$d(x, y) = \sup\{|x(t) - y(t)|: t \in A\}.$$
Then (X, d) is a metric space. The verification is similar to that for Example 7 except that we *cannot* assume that
$$\sup\{|x(t) - y(t)|: t \in A\} = |x(t_0) - y(t_0)|$$
for some $t_0 \in A$. Therefore the proof of $M3$ has to be revised as follows.

$M3$. For any $t_0 \in A$ we have
$$|x(t_0) - y(t_0)| \leqslant |x(t_0) - z(t_0)| + |z(t_0) - y(t_0)|$$
$$\leqslant \sup\{|x(t) - z(t)|: t \in A\}$$
$$+ \sup\{|z(t) - y(t)|: t \in A\}$$
$$= d(x, z) + d(z, y).$$
Therefore $d(x, z) + d(z, y)$ is an upper bound for the set $\{|x(t) - y(t)|: t \in A\}$ and so
$$d(x, y) = \sup\{|x(t) - y(t)|: t \in A\}$$
$$\leqslant d(x, z) + d(z, y).$$

Exercise 15 Let $x, y: [0, 1] \to \mathbb{R}$ be given by $x(t) = [t]$ (the integer part of t) and $y(t) = t^2$. Evaluate the distance between these functions using the 'sup' metric of Example 8.

9. Another way of measuring distance between continuous functions is by 'adding up' all the distances apart of their graphs. Let $X = C(a, b)$, the set of continuous functions $[a, b] \to \mathbb{R}$, and for $x, y \in X$ define $d(x, y)$ by
$$d(x, y) = \int_a^b |x(t) - y(t)| \, dt.$$
Then (X, d) is a metric space.

22   Metric spaces

Exercise 16 Show that the (X, d) defined in Example **9** does satisfy $M1$, $M2$ and $M3$. (You may assume all the necessary properties of integrals. In particular you will need the fact that, if x is continuous with $x(t) \geq 0$ for $t \in [a, b]$ and $\int_a^b x(t)\,dt = 0$, then $x(t) = 0$ for each t.)

We are now in a position to return to the question of whether a sequence of functions (or other objects) converges.

2.3 Sequences

A sequence is simply an infinite list. Examples of sequences are

$$1, 2, 4, 8, 16, \ldots$$

(which is a sequence – or at least the start of a sequence – of integers) and

$$(2, \tfrac{1}{2}), (3, \tfrac{1}{3}), (4, \tfrac{1}{4}), (5, \tfrac{1}{5}), \ldots$$

(which is a sequence of points of $\mathbb{R}^2$) and

$$i, \tfrac{1}{2} + \tfrac{1}{2}i, \tfrac{2}{3} + \tfrac{1}{3}i, \tfrac{3}{4} + \tfrac{1}{4}i, \ldots$$

(which is a sequence of members of $\mathbb{C}$). In general a *sequence* in a set X is a list

$$x_1, x_2, x_3, \ldots, x_n, \ldots$$

of members of X.

What do we mean by a converging sequence? We might guess that the sequence of real numbers

$$2, 1\tfrac{1}{2}, 1\tfrac{1}{3}, 1\tfrac{1}{4}, \ldots, 1 + \frac{1}{n}, \ldots$$

converges to 1. That is because the distances of the terms from 1 dwindles to nothing: the distance of the nth term from 1 (as always with respect to the usual metric on $\mathbb{R}$) is given by

$$|x_n - 1| = \left|\left(1 + \frac{1}{n}\right) - 1\right| = \frac{1}{n} \to 0 \quad \text{as } n \to \infty.$$

The sequence

$$\mathbf{x}_1 = (2\tfrac{1}{2}, 1),\ \mathbf{x}_2 = (2\tfrac{3}{4}, \tfrac{1}{2}),\ \mathbf{x}_3 = (2\tfrac{7}{8}, \tfrac{1}{3}), \ldots, \mathbf{x}_n = \left(3 - \frac{1}{2^n}, \frac{1}{n}\right), \ldots$$

in $\mathbb{R}^2$ would seem to be converging to $\mathbf{x} = (3, 0)$. To check this we look at the distance of the nth term from that $\mathbf{x}$:

Sequences

$$d(\mathbf{x}_n, \mathbf{x}) = d\left(\left(3 - \frac{1}{2^n}, \frac{1}{n}\right), (3, 0)\right)$$

$$= \left[\left(3 - \frac{1}{2^n} - 3\right)^2 + \left(\frac{1}{n} - 0\right)^2\right]^{1/2}$$

$$= \left(\frac{1}{4^n} + \frac{1}{n^2}\right)^{1/2}.$$

Again, this distance certainly tends to 0 as *n* gets large.

We can now see what we mean by a sequence converging in a metric space (X, d).

Definition In a metric space (X, d) the sequence $x_1, x_2, x_3, \ldots, x_n, \ldots$ of points of X *converges* to *limit* $x \in X$ if $d(x_n, x) \to 0$ (i.e. if the sequence $d(x_1, x), d(x_2, x), \ldots$ of **real numbers** converges to 0) as $n \to \infty$. We write $x_1, x_2, x_3, \ldots \to x$ or simply $x_n \to x$ as $n \to \infty$.

Exercise 17 Use the definition to show that the sequence

$$(0, 1), (\tfrac{1}{2}, \tfrac{5}{8}), (\tfrac{3}{4}, \tfrac{5}{9}), (\tfrac{7}{8}, \tfrac{17}{32}), \ldots, \left(1 - \frac{1}{2^{n-1}}, \frac{n^2 + 1}{2n^2}\right), \ldots$$

converges to $(1, \tfrac{1}{2})$ in $\mathbb{R}^2$.

Exercise 18 Show that the sequence

$$\cos \pi + i \sin \pi, \cos \frac{\pi}{2} + i \sin \frac{\pi}{2}, \cos \frac{\pi}{3} + i \sin \frac{\pi}{3},$$

$$\cos \frac{\pi}{4} + i \sin \frac{\pi}{4}, \ldots$$

converges to 1 in $\mathbb{C}$.

The sequences in these two exercises can be illustrated in the plane and, as far as intuition goes, they certainly seem to converge as claimed (Figure 2.5). It is also fairly clear that

$$(0, 1), (\tfrac{1}{2}, \tfrac{5}{8}), (\tfrac{3}{4}, \tfrac{5}{9}), (\tfrac{7}{8}, \tfrac{17}{32}), \ldots$$

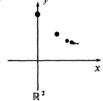
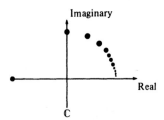

Fig. 2.5

24 Metric spaces

converges to $(1, \tfrac{1}{2})$ since the first coordinates tend to 1 and the second coordinates tend to $\tfrac{1}{2}$. And indeed it does turn out that this coordinate check is enough, as we now see.

Theorem 2.1 The sequence $(x_1, y_1), (x_2, y_2), (x_3, y_3), \ldots$ converges to (x, y) in $\mathbb{R}^2$ if and only if $x_1, x_2, x_3, \ldots$ converges to x and $y_1, y_2, y_3, \ldots$ converges to y in $\mathbb{R}$.

Proof Note firstly that for any (x, y) and $(x', y') \in \mathbb{R}^2$ we have
$$|x - x'| \leqslant [(x - x')^2 + (y - y')^2]^{1/2} = d((x, y), (x', y')).$$
So if $(x_1, y_1), (x_2, y_2), (x_3, y_3), \ldots \to (x, y)$ in $\mathbb{R}^2$, then
$$0 \leqslant |x_n - x| \leqslant d((x_n, y_n), (x, y)) \to 0 \quad \text{as } n \to \infty$$
and $|x_n - x| \to 0$ as $n \to \infty$. Hence $x_1, x_2, x_3, \ldots \to x$ and, similarly, $y_1, y_2, y_3, \ldots \to y$ in $\mathbb{R}$.

Conversely, if $x_1, x_2, x_3, \ldots \to x$ and $y_1, y_2, y_3, \ldots \to y$ in $\mathbb{R}$, then $|x_n - x| \to 0$ and $|y_n - y| \to 0$ as $n \to \infty$. Hence $(x_n - x)^2 \to 0$ and $(y_n - y)^2 \to 0$ as $n \to \infty$ and
$$d((x_n, y_n), (x, y)) = [(x_n - x)^2 + (y_n - y)^2]^{1/2} \to 0 \quad \text{as } n \to \infty.$$
Thus $(x_1, y_1), (x_2, y_2), (x_3, y_3), \ldots \to (x, y)$ in $\mathbb{R}^2$ as required. □

Exercise 19 Show that $(x_1, y_1, z_1), (x_2, y_2, z_2), (x_3, y_3, z_3), \ldots \to (x, y, z)$ in $\mathbb{R}^3$ if and only if $x_1, x_2, x_3, \ldots \to x$ and $y_1, y_2, y_3, \ldots \to y$ and $z_1, z_2, z_3, \ldots \to z$ in $\mathbb{R}$. If you are keen, state and prove the corresponding result for $\mathbb{R}^n$.

Exercise 20 Use the fact that the distance in $\mathbb{C}$ from $x_1 + x_2 i$ to $y_1 + y_2 i$ is the same as the distance in $\mathbb{R}^2$ from (x_1, x_2) to (y_1, y_2) to show that the sequence of complex numbers $z_1, z_2, z_3, \ldots$ converges to z if and only if $\operatorname{Re} z_1, \operatorname{Re} z_2, \operatorname{Re} z_3, \ldots$ converges to $\operatorname{Re} z$ and $\operatorname{Im} z_1, \operatorname{Im} z_2, \operatorname{Im} z_3, \ldots$ converges to $\operatorname{Im} z$. (Here, $\operatorname{Re} z$ denotes the real part of z and $\operatorname{Im} z$ denotes its imaginary part.)

All the above examples show that, as far as $\mathbb{R}, \mathbb{R}^2, \mathbb{R}^n$ and $\mathbb{C}$ are concerned, our abstract concept of convergence teaches us nothing that we could not have guessed anyway. It is only in the non-standard examples that intuition starts to let us down.

Exercise 21 Let $x_1, x_2, x_3, \ldots \to x$ in a set X with the discrete metric d. (So $d(x_n, x)$ is either 0 or 1 for each n.) Show that, apart from a finite number of the x_ns, all the terms of the

Sequences

sequence are *equal* to x. (Hint: $d(x_n, x)$ must eventually be less than 1.)

Before proceeding note that the sequence $(2\frac{1}{2}, \frac{1}{2})$, $(2\frac{1}{4}, \frac{1}{4})$, $(2\frac{1}{8}, \frac{1}{8})$, ... in $\mathbb{R}^2$ converges to (2, 0) with respect to the usual metric but it does not converge with respect to the discrete metric. So when talking about convergence we ought to make it clear which metric is involved. As mentioned earlier, unless we say otherwise we assume that the usual metric is being used.

Now let us turn to the function spaces. Recall that $C(1, 2)$ is the set of continuous functions $[1, 2] \to \mathbb{R}$ and that the metric which will concern us most on this set is the 'max' metric. In this space consider the sequence $x_1, x_2, x_3, \ldots$ given by

$$x_1(t) = \frac{t}{1+t}, x_2(t) = \frac{2t}{2+t}, x_3(t) = \frac{3t}{3+t}, \ldots, x_n(t) = \frac{nt}{n+t}, \ldots$$

Figure 2.6 illustrates the graphs of these functions. The graphs seem

Fig. 2.6

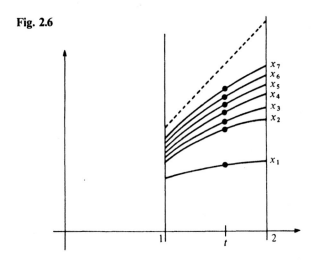

to suggest that this sequence of functions converges to the function x given by $x(t) = t$. Let us check this:

$$0 \leqslant d(x_n, x) = \max\{|x_n(t) - x(t)|: 1 \leqslant t \leqslant 2\}$$

$$= \max\left\{\left|\frac{nt}{n+t} - t\right|: 1 \leqslant t \leqslant 2\right\}$$

$$= \max\left\{\left|\frac{nt - t(n+t)}{n+t}\right| : 1 \leqslant t \leqslant 2\right\}$$

$$= \max\left\{\frac{t^2}{n+t} : 1 \leqslant t \leqslant 2\right\}$$

$$< \max\left\{\frac{t^2}{n} : 1 \leqslant t \leqslant 2\right\} = \frac{4}{n} \to 0 \quad \text{as } n \to \infty.$$

Hence $d(x_n, x) \to 0$ as $n \to \infty$ and we have checked formally that $x_1, x_2, x_3, \ldots$ does converge to x in $C(1, 2)$. How could we have guessed the limit without drawing the graphs? Fix t and consider the values of $x_1, x_2, x_3, \ldots$ at t, namely

$$\frac{t}{1+t}, \frac{2t}{2+t}, \frac{3t}{3+t}, \ldots, \frac{nt}{n+t}, \ldots.$$

Since t is fixed this is merely a sequence of real numbers and (since $nt/(n+t) = t/(1+t/n)$) it converges to t. So for each fixed coordinate the functions converge to t, as shown in the above figure: that leads us to try x given by $x(t) = t$ as the limit. This is justified in general by the following result:

Theorem 2.2 Let $x, x_1, x_2, x_3, \ldots$ be functions in $C(a, b)$ such that $x_1, x_2, x_3, \ldots \to x$. Then, for any $t \in [a, b]$, $x_1(t), x_2(t), x_3(t), \ldots \to x(t)$ in $\mathbb{R}$.

Proof Since $d(x_n, x)$ is the maximum of all the $|x_n(t) - x(t)|$ for $t \in [a, b]$ it follows that, for any particular t,

$$|x_n(t) - x(t)| \leqslant d(x_n, x).$$

So if $d(x_n, x) \to 0$ as $n \to \infty$, then $|x_n(t) - x(t)| \to 0$ as $n \to \infty$; i.e. if $x_1, x_2, x_3, \ldots \to x$ in $C(a, b)$, then $x_1(t), x_2(t), x_3(t), \ldots \to x(t)$ in $\mathbb{R}$. □

So the moral of that result is that if we are trying to decide whether a sequence $x_1, x_2, x_3, \ldots$ in $C(a, b)$ converges, then for each $t \in [a, b]$ find the limit of the real sequence $x_1(t), x_2(t), x_3(t), \ldots$. (If it fails to exist for some t, then the above theorem shows that $x_1, x_2, x_3, \ldots$ does not converge in $C(a, b)$.) If the limit exists denote it by $x(t)$. The collection of these $x(t)$s for $t \in [a, b]$ defines a new function $x: [a, b] \to \mathbb{R}$ and we check whether $x \in C(a, b)$ and whether it is the limit of $x_1, x_2, x_3, \ldots$.

Exercise 22 Let $x_1, x_2, x_3, \ldots$ in $C(0, 1)$ be defined by $x_1(t) = t$, $x_2(t) = t^2, \ldots, x_n(t) = t^n, \ldots$. Find the limit of the real

Sequences

sequence $x_1(t), x_2(t), x_3(t), \ldots$ for each $t \in [0, 1]$, and denote this limit by $x(t)$. Show that the function defined in this way is not continuous and is not, therefore, in $C(0, 1)$. (Hence $x_1, x_2, x_3, \ldots$ does not converge in $C(0, 1)$.)

Exercise 23 Let $x_1, x_2, x_3, \ldots$ in $C(0, 1)$ be defined by

$$x_1(t) = \frac{t}{1+t^2}, x_2(t) = \frac{2t}{1+4t^2}, \ldots, x_n(t) = \frac{nt}{1+n^2t^2}, \ldots.$$

Show that $x_1(t), x_2(t), x_3(t), \ldots \to 0$ for each $t \in [0, 1]$. (Hence, by the above theorem, if $x_1, x_2, x_3, \ldots \to x$ then $x(t) = 0$ for each $t \in [0, 1]$.) But show that, for the usual 'max' metric on $C(0, 1)$, if x is the zero function then $d(x_n, x) = \frac{1}{2}$. Hence confirm that $x_1, x_2, x_3, \ldots$ does not converge in $C(0, 1)$.

So, by the above theorem, convergence in $C(a, b)$ implies 'coordinatewise' convergence of the functions. But, as Exercises 22 and 23 show, coordinatewise convergence is not enough to ensure convergence in $C(a, b)$. Coordinates merely enable us to define the coordinatewise limit x and then we must check whether x is in $C(a, b)$ and is also the limit with respect to the 'max' metric. The sequence of functions in Exercise 23 is illustrated in Figure 2.7.

It is clear that, in some way, the sequence of functions illustrated below does not settle down to the zero function as neatly as the earlier illustrated sequence settled down to the function $x(t) = t$. Convergence in $C(a, b)$ with respect to the 'max' metric requires much more

Fig. 2.7

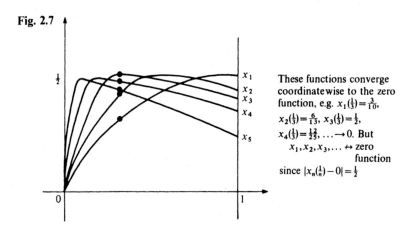

These functions converge coordinatewise to the zero function, e.g. $x_1(\frac{1}{3}) = \frac{3}{10}$, $x_2(\frac{1}{3}) = \frac{6}{13}$, $x_3(\frac{1}{3}) = \frac{1}{2}$, $x_4(\frac{1}{3}) = \frac{12}{25}, \ldots \to 0$. But $x_1, x_2, x_3, \ldots \not\to$ zero function since $|x_n(\frac{1}{n}) - 0| = \frac{1}{2}$

28 Metric spaces

than coordinatewise convergence. Readers interested in seeing this stronger concept in more traditional approaches elsewhere will find it referred to as *uniform convergence*.

We have now set up all the required machinery for asking whether sequences converge. Before proceeding we have two pure mathematical points to make to conclude this chapter. The first is that, given a sequence $x_1, x_2, x_3, \ldots$, a *subsequence* of it is another sequence $x_{k_1}, x_{k_2}, x_{k_3}, \ldots$ obtained from the original by picking out the k_1st, k_2nd, k_3rd terms, etc., where $k_1 < k_2 < k_3 < \ldots$. If a sequence converges in a metric space then any subsequence of it converges with the same limit. For if $d(x_n, x) \to 0$ as $n \to \infty$, then $d(x_{k_n}, x)$ (being $d(x_m, x)$ for some $m \geqslant n$) $\to 0$ as $n \to \infty$. The second point we make is that we can talk unambiguously of *the* limit of a convergent sequence in a metric space, as we now confirm.

Theorem 2.3 If $x_1, x_2, x_3, \ldots \to x$ and $x_1, x_2, x_3, \ldots \to y$ in (X, d), then $x = y$.

Proof If $d(x_n, x) \to 0$ and $d(x_n, y) \to 0$ as $n \to \infty$, then
$$0 \leqslant d(x, y) \leqslant d(x, x_n) + d(x_n, y) \to 0$$
and so $d(x, y)$ must equal 0: hence $x = y$. □

This result shows that a sequence can have at most one limit and so we can talk unambiguously of *the* limit of a convergent sequence. The proof is also a neat way of concluding this chapter, for it shows that by our abstract approach we can give transparent proofs which work for *all* metric spaces and which are no harder than the corresponding proofs for $\mathbb{R}$ alone.

3

The three Cs

3.1 Iteration revisited

Let us return briefly to the idea of solving a real equation of the form $x = f(x)$ by iterating with the function f. But now let us try to solve the equation with the additional constraint that the root should lie in some given set.

For example let us try to find a root of $x^3 - 10x^2 + 31x - 30 = 0$ with $x > 2$. We will rearrange the equation as

$$x = \frac{-x^3 + 10x^2 + 30}{31} = f(x)$$

and iterate with f starting at $x_1 = 2.5$ (which does satisfy $x_1 > 2$). Then $x_2 = f(x_1) \approx 2.4798$, $x_3 = f(x_2) \approx 2.4595$, $x_4 = f(x_3) \approx 2.4392$, and so on. In this way we get a convergent sequence, so its limit is a root of the given equation. Also the terms of the sequence all satisfy $x_n > 2$. So far, so good. But unfortunately the limit of the sequence turns out to be 2 – which does *not* satisfy the condition $x > 2$.

As another example let us try to find a rational (i.e. fractional) root of $x^4 + 2x^3 - 3x^2 - 4x + 2 = 0$. We will rearrange this as

$$x = \frac{x^4 + 2x^3 - 3x^2 + 2}{4} = f(x)$$

and iterate with f starting with $x_1 = 1$. This time, as we are looking for a rational root, we will write the terms of the sequence as fractions. We find that $x_2 = f(x_1) = \frac{1}{2}$, $x_3 = f(x_2) = \frac{25}{64}$, $x_4 = f(x_3) = \frac{28265057}{67108864}$, and so on. In this way we get a convergent sequence so its limit is a root of the given equation, and all the terms of the sequence are rationals. So far, so good. But the limit of the sequence turns out to be $\sqrt{2} - 1$, which is not rational, so again this technique has failed to find a root with the additional constraint.

In the first example we were trying to find a root in the open interval $]2, \infty[$: we found a sequence in the required set converging to a root, but unfortunately the limit was outside the set. In the second example we were trying to find a root in the set of rationals: we found a sequence in the required set converging to a root, but unfortunately the limit was again outside the set. For this method to work, enabling us to find a root in the set A, say, we must first ensure that A has the property that for convergent sequences in A their limits are also in A. This rules out open intervals like $]0, 1[$ or $]2, \infty[$, but not closed ones. We will look now at the corresponding property of a set A in an arbitrary metric space: this brings us to closed sets, the first of our three Cs.

3.2   Closed sets

Definition In a metric space (X, d) the set $A \subseteq X$ is *closed* if whenever $a_1, a_2, a_3, \ldots \in A$ and $a_1, a_2, a_3, \ldots \to a$ it follows that $a \in A$.

So, for example, the interval $[0, 1]$ is closed in $\mathbb{R}$. For if $a_1, a_2, a_3, \ldots \in [0, 1]$ and $a_1, a_2, a_3, \ldots \to a$, then $a_1 \geqslant 0, a_2 \geqslant 0, a_3 \geqslant 0, \ldots$ and so $a \geqslant 0$ (by a straightforward property of real sequences), and $a_1 \leqslant 1$, $a_2 \leqslant 1, a_3 \leqslant 1, \ldots$ and so $a \leqslant 1$; hence $a \in [0, 1]$. Similarly, in $\mathbb{R}$ all closed intervals are closed (surprise, surprise). However, the interval $]0, 1[$ is not closed: e.g. $\frac{1}{2}, \frac{1}{3}, \frac{1}{4}, \ldots$ are all in that set but their limit, 0, is not. The set $\mathbb{Q}$ of rationals is not closed in $\mathbb{R}$, for, as we have seen, it is possible to have a sequence of rationals converging to an irrational limit.

Exercise 24 Let (X, d) be a metric space and let $x \in X$. Show that the set $\{x\}$ is closed.

Exercise 25 Show that the set of irrational numbers is not closed in $\mathbb{R}$.

Now let us look at some subsets of $\mathbb{R}^2$. Consider first the set $A = \{(x, y): x^2 + y^2 < 5\}$. The sequence $(0, 1), (\frac{1}{2}, 1\frac{1}{2}), (\frac{2}{3}, 1\frac{2}{3}), (\frac{3}{4}, 1\frac{3}{4}), \ldots$ of points of A converges but its limit $(1, 2)$ is not in A since $1^2 + 2^2$ *equals* 5. Hence A is not closed. On the other hand, the set $B = \{(x, y): x^2 + y^2 \leqslant 5\}$ *is* closed. For if $(x_1, y_1), (x_2, y_2), (x_3, y_3), \ldots$ are points of B converging to (x, y), then

$$x_1^2 + y_1^2, x_2^2 + y_2^2, x_3^2 + y_3^2, \ldots \to x^2 + y^2$$
$$\leqslant 5 \quad \leqslant 5 \quad \leqslant 5$$

Closed sets

Fig. 3.1

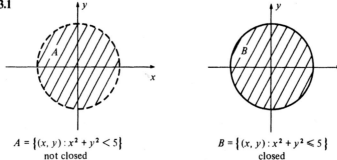

$A = \{(x, y) : x^2 + y^2 < 5\}$
not closed

$B = \{(x, y) : x^2 + y^2 \leq 5\}$
closed

and so $x^2 + y^2 \leq 5$. Hence the limit (x, y) is in B and the set B is closed. These sets are illustrated in Figure 3.1.

Another way of writing B is as $f^{-1}(]-\infty, 5])$ where $f: \mathbb{R}^2 \to \mathbb{R}$ is given by $f(x, y) = x^2 + y^2$. In other words, B consists of all those (x, y) for which $f(x, y)$ lies in $]-\infty, 5]$. Our verification above that B was closed depended upon two facts. Firstly we used the property that if

$$(x_1, y_1), (x_2, y_2), (x_3, y_3), \ldots \to (x, y),$$

then

$$f(x_1, y_1) = x_1^2 + y_1^2, f(x_2, y_2) = x_2^2 + y_2^2,$$
$$f(x_3, y_3) = x_3^2 + y_3^2, \ldots \to f(x, y) = x^2 + y^2.$$

Secondly we used the fact that if a convergent sequence lies in $]-\infty, 5]$, then so does its limit. The former property is simply the continuity of f (informally, that if (x', y') is close to (x, y) then $f(x', y')$ is close to $f(x, y)$) and the latter property is, of course, simply that $]-\infty, 5]$ is closed.

Our next theorem will generalise that result and enable us to find many closed sets in $\mathbb{R}^2$. For this we will need the concept of a continuous function f of two variables. A function $f: \mathbb{R}^2 \to \mathbb{R}$ is *continuous* if whenever

$$(x_1, y_1), (x_2, y_2), (x_3, y_3), \ldots \to (x, y)$$

it follows that

$$f(x_1, y_1), f(x_2, y_2), f(x_3, y_3), \ldots \to f(x, y).$$

(This formal property need not concern us unduly: most uncontrived functions $\mathbb{R}^2 \to \mathbb{R}$ are continuous.)

Theorem 3.1 Let $f: \mathbb{R}^2 \to \mathbb{R}$ be continuous and let A be closed in $\mathbb{R}$. Then $f^{-1}(A)$ $(= \{(x, y): f(x, y) \in A\})$ is closed in $\mathbb{R}^2$.

Proof We must consider a convergent sequence in $f^{-1}(A)$ and show that the limit is also in $f^{-1}(A)$. So let $(x_1, y_1), (x_2, y_2), (x_3, y_3), \ldots$ be in $f^{-1}(A)$ with

$$\underbrace{(x_1, y_1), (x_2, y_2), (x_3, y_3), \ldots}_{\in f^{-1}(A)} \to (x, y).$$

Then, by the continuity of f,

$$\underbrace{f(x_1, y_1), f(x_2, y_2), f(x_3, y_3), \ldots}_{\in A} \to f(x, y).$$

But A is a closed set and therefore it contains the limit $f(x, y)$. Hence $(x, y) \in f^{-1}(A)$ and we have shown that every convergent sequence in $f^{-1}(A)$ has its limit in $f^{-1}(A)$; i.e. $f^{-1}(A)$ is closed as claimed. □

Exercise 26 Show that the set $\{(x, y): xy \geq 1\}$ is closed in $\mathbb{R}^2$.

Exercise 27 Show that all straight lines in $\mathbb{R}^2$ are closed. (You may recall from Exercise 24 that the set $\{0\}$ is closed in $\mathbb{R}$ and then express a straight line in the form $f^{-1}(\{0\}) = \{(x, y): f(x, y) = 0\}$ for some suitably chosen continuous function f.)

Exercise 28 Let $x_1, x_2, x_3, \ldots \to x$ in $C(a, b)$ and let $x_1, x_2, x_3, \ldots$ all be functions from $[a, b]$ to $[c, d]$. Use Theorem 2.2 to show that, for each $t \in [a, b]$, $x(t) \in [c, d]$. Deduce that the set of continuous functions $[a, b] \to [c, d]$ is a closed set in $C(a, b)$.

Theorem 3.1 is a special case of a very general result concerning continuous functions and closed sets. The general result need not concern the reader only interested in our central theme of iteration, but for those readers who wish to use metric spaces to obtain a better understanding of analysis a fuller version of Theorem 3.1 will appear in the final chapter.

Exercise 29 Show that the quadrant $\{(x, y): x^2 + y^2 \leq 1, x \geq 0$ and $y \geq 0\}$ is closed in $\mathbb{R}^2$.

There are various ways of tackling Exercise 29, one being from the basic definition. In that case let $(x_1, y_1), (x_2, y_2), (x_3, y_3), \ldots \to (x, y)$ with each (x_n, y_n) satisfying $x_n^2 + y_n^2 \leq 1$, $x_n \geq 0$ and $y_n \geq 0$ and deduce that $x^2 + y^2 \leq 1$, $x \geq 0$ and $y \geq 0$. A more far-reaching way of showing

Closed sets

that the given quadrant is closed is to note that
$$\{(x, y): x^2 + y^2 \leq 1, x \geq 0 \text{ and } y \geq 0\}$$
$$= \{(x, y): x^2 + y^2 \leq 1\} \cap \{(x, y): x \geq 0\} \cap \{(x, y): y \geq 0\}.$$
Each of the three sets on the right is closed (by Theorem 3.1) and, as we see now, any intersection of closed sets is closed.

Theorem 3.2 Any intersection of closed sets in a metric space is itself closed.

Proof Let A be the intersection of any number of closed sets in a metric space and let $a_1, a_2, a_3, \ldots$ be points of A with $a_1, a_2, a_3, \ldots \to a$. Then to show that A is closed it remains to prove that $a \in A$. But since $a_1, a_2, a_3, \ldots$ are in the intersection of all the closed sets in question, they are in each of the closed sets. So the limit a is also in each of the closed sets. Hence a is in the intersection of all these closed sets, namely A. So we have shown that every convergent sequence in A has its limit in A; i.e. that A is closed. □

Exercise 30 A *half-plane* in $\mathbb{R}^2$ is a set of the form $\{(x, y): ax + by \leq c\}$ where a, b and c are real constants with not both a and b zero. Show that half-planes are closed in $\mathbb{R}^2$ and deduce that all triangles are closed in $\mathbb{R}^2$. The keen reader might like to deduce that all polygons are closed.

Exercise 31 Let $A \subseteq C(0, 1)$ consist of those functions x in $C(0, 1)$ with $x(0) = 0$. Use Theorem 2.2 to show that if $x_1, x_2, x_3, \ldots$ are in A and $x_1, x_2, x_3, \ldots \to x$, then $x \in A$; i.e. that A is closed. Deduce that the set
$$\{x \in C(0, 1): x(0) = 0, x(\tfrac{1}{100}) = 1, x(\tfrac{2}{100}) = 2, \ldots,$$
$$x(\tfrac{99}{100}) = 99 \text{ and } x(1) = 100\}$$
is closed.

Now let us look at a problem similar to that in Exercise 29 but with 'and' replaced by 'or'. Let us try to show that the set
$$\{(x, y): x^2 + y^2 \leq 1 \text{ or } x \geq 0 \text{ or } y \geq 0\}$$
is closed in $\mathbb{R}^2$ (Figure 3.2). Consider any sequence $(x_1, y_1), (x_2, y_2), (x_3, y_3), \ldots$ in this set with $(x_1, y_1), (x_2, y_2), (x_3, y_3), \ldots \to (x, y)$. Since there is an infinite number of terms in the sequence and each satisfies at least one of the three conditions $x_n^2 + y_n^2 \leq 1$, $x_n \geq 0$ or $y_n \geq 0$, we must be able to pick out an infinite number of the (x_n, y_n)s which all

Fig. 3.2

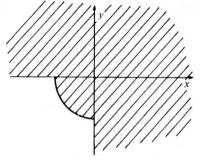

satisfy the same condition. (Imagine sorting the (x_1, y_1), (x_2, y_2), $(x_3, y_3), \ldots$ into three piles: the first pile consists of those satisfying $x_n^2 + y_n^2 \leqslant 1$, the second pile consists of those of the remaining ones satisfying $x_n \geqslant 0$, and the third pile is all the rest – which must satisfy $y_n \geqslant 0$. At least one of these piles must be infinite.) Assume for example that an infinite number of the terms satisfy $x_n^2 + y_n^2 \leqslant 1$; pick out a subsequence with this property: $(x_{k_1}, y_{k_1}), (x_{k_2}, y_{k_2}), (x_{k_3}, y_{k_3}), \ldots,$ say. This sequence also converges to (x, y) and so this limit satisfies $x^2 + y^2 \leqslant 1$: hence (x, y) is in the set

$$\{(x, y): x^2 + y^2 \leqslant 1 \text{ or } x \geqslant 0 \text{ or } y \geqslant 0\}.$$

A similar argument would have worked if some subsequence had satisfied one of the other conditions. So, in general, any convergent sequence in the given set has its limit in the set; i.e.

$$\{(x, y): x^2 + y^2 \leqslant 1 \text{ or } x \geqslant 0 \text{ or } y \geqslant 0\} \text{ is closed.}$$

Again, that is only a particular case of the following general result.

Theorem 3.3 Let $A_1, A_2, \ldots, A_k$ be closed sets in a metric space. Then $A_1 \cup A_2 \cup \ldots \cup A_k$ is closed.

Proof Let $A = A_1 \cup A_2 \cup \ldots \cup A_k$ and let $a_1, a_2, a_3, \ldots$ be in A with $a_1, a_2, a_3, \ldots \to a$. We must show that $a \in A$. But some A_i must contain an infinite number of the $a_1, a_2, a_3, \ldots$ (otherwise $A_1, A_2, \ldots, A_k$ would only contain in total a finite number of the $a_1, a_2, a_3, \ldots$). So assume that the subsequence $a_{k_1}, a_{k_2}, a_{k_3}, \ldots$ lies in A_i (which we know is closed). Also $a_{k_1}, a_{k_2}, a_{k_3}, \ldots \to a$ and so $a \in A_i$. Hence $a \in A_1 \cup A_2 \cup \ldots \cup A_k = A$ and we have shown that every convergent sequence in A has its limit in A; i.e. A is closed. □

Closed sets

Exercise 32 Show that the set
$\{(x, y): y = nx \text{ for some integer } n \text{ between 1 and 100}\}$
is closed in $\mathbb{R}^2$.

Exercise 33 Show that the set
$\{x \in C(0, 100): x(t) = t \text{ for some integer } t\}$
is closed in $C(0, 100)$.

In each of those exercises the given set could be expressed as a union of a finite number of sets which we already know to be closed (from Exercise 27 and the first part of Exercise 31).

The reader may have noticed a subtle difference between Theorems 3.2 and 3.3: the former concerned *any* intersection of closed sets, but the latter only concerned unions of a *finite number* of closed sets. The following exercise shows that Theorem 3.3 cannot be extended to arbitrary unions of closed sets.

Exercise 34 Let $A \subseteq \mathbb{R}^2$ be given by
$A = \{(x, y): y = nx \text{ for some integer } n\}$.
Show that A is a union of straight lines (and hence of closed sets). But show that A is not closed by finding a sequence of points in A converging to $(0, 1)$ (which is not in A).

That concludes our basic introduction to closed sets. We have only included enough to give the reader a general idea of the concept as far as it is needed in our subsequent applications to iterative processes. In Chapter 5, our optional pure mathematical chapter, we shall return to this concept and see its role in analysis.

The idea of a closed set was a most natural one in our approach via iteration and sequences. But if we had approached the subject of metric spaces from a different standpoint (perhaps by generalising ε–δ arguments from real analysis, or as a preparation for the even more general subject of 'topology') then we might have more naturally come across the concept of an 'open' set. Open sets are precisely the complements of closed sets. There is no real need for both concepts because all theorems about open sets can be stated in terms of closed sets. So we shall restrict attention to closed sets for the moment, and return to open sets in the final chapter.

3.3 An internal test for convergence

Given a sequence is it possible to tell by looking at its terms whether it converges or not? In other words, can we find a test for convergence which does not need prior knowledge of the limit?

The real sequence

$$\tfrac{1}{2}, \tfrac{3}{4}, \tfrac{7}{8}, \tfrac{15}{16}, \ldots$$

certainly seems to be converging, but then perhaps so does

$$1,\ 1+\tfrac{1}{2},\ 1+\tfrac{1}{2}+\tfrac{1}{3},\ 1+\tfrac{1}{2}+\tfrac{1}{3}+\tfrac{1}{4}, \ldots.$$

Is there some test which we can apply to these terms to decide whether they converge or not? Similarly, in $C(0, 1)$ given the sequence $x_1, x_2, x_3, \ldots$ defined by

$$x_1(t) = \frac{t}{1+t}, \quad x_2(t) = \frac{2t}{2+t}, \quad x_3(t) = \frac{3t}{3+t}, \quad \ldots$$

or the sequence $y_1, y_2, y_3, \ldots$ defined by

$$y_1(t) = \frac{t}{1+t^2}, \quad y_2(t) = \frac{2t}{1+4t^2}, \quad y_3(t) = \frac{3t}{1+9t^2}, \quad \ldots$$

is there some way of looking at the terms and deciding whether the sequences are convergent?

The only method available to us so far is to guess the limit and then check that the terms do get arbitrarily close to that limit. For example, we would guess that

$$x_1 = \tfrac{1}{2}, \quad x_2 = \tfrac{3}{4}, \quad x_3 = \tfrac{7}{8}, \quad \ldots$$

converges to 1; we would then check formally that

$$|x_n - 1| = \left|\left(1 - \frac{1}{2^n}\right) - 1\right| = \frac{1}{2^n} \to 0 \quad \text{as } n \to \infty.$$

But perhaps we can test the convergence of a sequence by looking at the distances apart of the terms in the sequence without ever having to consider the limit at all. In this example for $m \geqslant n$ the distance between terms is given by

$$|x_m - x_n| = \left|\left(1 - \frac{1}{2^m}\right) - \left(1 - \frac{1}{2^n}\right)\right| = \frac{1}{2^n} - \frac{1}{2^m}$$

which tends to zero as m and n get large. However, for the divergent sequence

$$y_1 = 1, \quad y_2 = 1 + \tfrac{1}{2}, \quad y_3 = 1 + \tfrac{1}{2} + \tfrac{1}{3}, \quad \ldots$$

the distance from the nth to the $2n$th term is given by

An internal test for convergence

$$|y_{2n} - y_n| = \left|\left(1 + \frac{1}{2} + \frac{1}{3} + \cdots + \frac{1}{2n}\right) - \left(1 + \frac{1}{2} + \frac{1}{3} + \cdots + \frac{1}{n}\right)\right|$$

$$= \frac{1}{n+1} + \frac{1}{n+2} + \cdots + \frac{1}{2n} \geq \underbrace{\frac{1}{2n} + \cdots + \frac{1}{2n}}_{n \text{ times}} = \frac{1}{2};$$

and so here it is not the case that $|y_m - y_n| \to 0$ as m and n get large.

In $C(0, 1)$ the convergent sequence $x_1, x_2, x_3, \ldots$ given by

$$x_1(t) = \frac{t}{1+t}, \quad x_2(t) = \frac{2t}{2+t}, \quad x_3(t) = \frac{3t}{3+t}, \quad \ldots$$

also has the property that the distance between the mth and nth terms tends to zero as m and n get large. For if $m \geq n$ then

$$d(x_m, x_n) = \left|\frac{mt_0}{m+t_0} - \frac{nt_0}{n+t_0}\right| \quad \text{for some } t_0 \in [0, 1]$$

$$= \left|\frac{(m-n)t_0^2}{(m+t_0)(n+t_0)}\right| \leq \frac{t_0^2}{n+t_0} \leq \frac{1}{n} \to 0.$$

However, as we saw in Chapter 2, the sequence given by

$$y_1(t) = \frac{t}{1+t^2}, \quad y_2(t) = \frac{2t}{1+4t^2}, \quad y_3(t) = \frac{3t}{1+9t^2}, \quad \ldots$$

diverges in $C(0, 1)$, and it turns out that this sequence fails to have the property that $d(y_m, y_n) \to 0$ as $m, n \to \infty$. For

$$d(y_{3n}, y_n) = \max \{|y_{3n}(t) - y_n(t)| : 0 \leq t \leq 1\}$$

$$\geq \left|y_{3n}\left(\frac{1}{n}\right) - y_n\left(\frac{1}{n}\right)\right|$$

$$= \left|\frac{3n \cdot 1/n}{1 + (3n)^2 (1/n)^2} - \frac{n \cdot 1/n}{1 + n^2 (1/n)^2}\right| = |\tfrac{3}{10} - \tfrac{1}{2}| = \tfrac{1}{5}.$$

So perhaps we have reason to suspect that a sequence $x_1, x_2, x_3, \ldots$ converges in a metric space (X, d) if and only if $d(x_m, x_n) \to 0$ as $m, n \to \infty$. Certainly convergent sequences do have this property, as we now see.

Theorem 3.4 If $x_1, x_2, x_3, \ldots$ is a convergent sequence in a metric space (X, d), then $d(x_m, x_n) \to 0$ as $m, n \to \infty$.

Proof Assume that the limit of the sequence is x. Then

$$0 \leq d(x_m, x_n) \leq d(x_m, x) + d(x, x_n) \to 0 + 0 = 0 \quad \text{as } m, n \to \infty.$$

Hence $d(x_m, x_n) \to 0$ as $m, n \to \infty$. $\square$

38   The three Cs

Exercise 35 Let $x_1, x_2, x_3, \ldots$ in $C(0, 1)$ be given by
$$x_1(t) = \frac{t^2}{1+t^2}, \quad x_2(t) = \frac{2t^2}{1+2t^2}, \quad x_3(t) = \frac{3t^2}{1+3t^2}, \quad \ldots$$
By considering $t = 1/n$ show that $d(x_{n^2}, x_{2n^2}) \geq \frac{1}{6}$ and use Theorem 3.4 to deduce that this sequence is not convergent.

What about the converse of Theorem 3.4? If a sequence $x_1, x_2, x_3, \ldots$ in a metric space (X, d) has the property that $d(x_m, x_n) \to 0$ as $m, n \to \infty$, can we be sure that the sequence converges? Consider the metric space (X, d) where X is the set of rationals and d is the usual real distance. Let $x_1, x_2, x_3, \ldots$ be given by

$$x_1 = 1 \quad \text{and} \quad x_{n+1} = \frac{1}{x_n} + \frac{x_n}{2} \quad \text{for } n \geq 1.$$

Then
$$x_1 = 1, \quad x_2 = \tfrac{1}{1} + \tfrac{1}{2} = \tfrac{3}{2}, \quad x_3 = \tfrac{2}{3} + \tfrac{3}{4} = \tfrac{17}{12}, \quad x_4 = \tfrac{577}{408}, \quad \ldots$$
and so we get a sequence of real numbers. This is quite a well-known sequence used to find $\sqrt{2}$ by iterative techniques for, as we can check with a calculator, this sequence seems to converge to $\sqrt{2}$. So, by Theorem 3.4, $d(x_m, x_n) = |x_m - x_n| \to 0$ as $m, n \to \infty$. But now we are interested in X, the set of rational numbers. The members of the sequence are all rationals and so we have a sequence in X with $d(x_m, x_n) \to 0$ as $m, n \to \infty$. However, the sequence *does not converge in X* since there is no $x \in X$ with $x_1, x_2, x_3, \ldots \to x$. This example may seem a little contrived, but in general a metric space (X, d) may consist of a set like the rationals which can have a sequence $x_1, x_2, x_3, \ldots$ with $d(x_m, x_n) \to 0$ as $m, n \to \infty$ for which there exists no $x \in X$ with $x_1, x_2, x_3, \ldots \to x$.

In our results concerning iteration in a metric space we are not going to be able to allow our spaces to have 'missing limits' in that sense; and this idea leads us on to the second of our three Cs.

3.4   Complete sets

Definition The set $A \subseteq X$ is *complete* in the metric space (X, d) if whenever $a_1, a_2, a_3, \ldots$ is a sequence in A with $d(a_m, a_n) \to 0$ as $m, n \to \infty$, it follows that the sequence is convergent *with its limit in A*.

So the complete sets are precisely those for which the $d(a_m, a_n) \to 0$ condition *is* sufficient to ensure convergence. Our above observations

Complete sets

show that $\mathbb{Q}$, the set of rationals, is not complete in $\mathbb{R}$. But we will now see that $\mathbb{R}$ itself is complete (with the usual metrics). This is the one point in our development of the subject where we ought to pause to question the assumptions which we make about $\mathbb{R}$.

A sequence $x_1, x_2, x_3, \ldots$ is *bounded above* if there exists a number u with $x_n \leqslant u$ for all n. Similarly, it is *bounded below* if there exists a number l with $l \leqslant x_n$ for all n. It is *bounded* if it is bounded above and below. In other words, naturally enough, a sequence is bounded if it lies entirely between two 'bounds'. A common assumption in real analysis is that a sequence which is increasing and which is bounded above must be convergent. From this it can be proved easily that a bounded sequence has a convergent subsequence. (The reader interested in these results can find proofs, for example, in Copson's or Sutherland's books listed on page 103.) So we shall assume without proof that any bounded real sequence has a convergent subsequence. We are now able to deduce the completeness of $\mathbb{R}$.

Theorem 3.5 $\mathbb{R}$ is complete.

Proof Let $a_1, a_2, a_3, \ldots$ be a sequence in $\mathbb{R}$ with the property that $|a_m - a_n| \to 0$ as $m, n \to \infty$. Then in particular there exists a positive integer N with $|a_m - a_n| < 1$ for $m, n \geqslant N$. But then the whole sequence lies in the set

$$\{a_1, a_2, \ldots, a_{N-1}\} \cup [a_N - 1, a_N + 1]$$

and so the sequence is bounded. Hence there is a convergent subsequence $a_{k_1}, a_{k_2}, a_{k_3}, \ldots \to a$, say. But then

$$0 \leqslant |a_n - a| \leqslant |a_n - a_{k_n}| + |a_{k_n} - a|$$

$$\downarrow \qquad \qquad \downarrow$$
$$0 \qquad \qquad 0$$

since $|a_n - a_m| \to 0$ for large n and m; since $a_{k_n} \to a$

and so $|a_n - a| \to 0$ as $n \to \infty$; i.e. $a_1, a_2, a_3, \ldots \to a$. Therefore the condition $|a_m - a_n| \to 0$ as $m, n \to \infty$ ensures the convergence of $a_1, a_2, a_3, \ldots$ in $\mathbb{R}$; i.e. $\mathbb{R}$ is complete. □

Sequences for which $d(a_m, a_n) \to 0$ as $m, n \to \infty$ are often referred to as *Cauchy sequences* and Theorem 3.5 will be found in books on real analysis as the 'Cauchy criterion for convergence'.

Which subsets of $\mathbb{R}$ are complete? The interval $[0,1]$ is complete. To see this let $a_1, a_2, a_3, \ldots$ be in $[0,1]$ with the property that $|a_m - a_n| \to 0$ as $m, n \to \infty$. But then, of course, $a_1, a_2, a_3, \ldots$ are in $\mathbb{R}$ and $|a_m - a_n| \to 0$ as $m, n \to \infty$. Therefore, by Theorem 3.5, the sequence is convergent in $\mathbb{R}$: $a_1, a_2, a_3, \ldots \to a$, say. But since $[0,1]$ is closed and contains $a_1, a_2, a_3, \ldots$ it also contains the limit a. So whenever $a_1, a_2, a_3, \ldots$ are in $[0,1]$ and $|a_m - a_n| \to 0$ as $m, n \to \infty$ it follows that the sequence converges to a limit in $[0,1]$; i.e. $[0,1]$ is complete.

The reader will probably see that we have only used the completeness of $\mathbb{R}$ and the closedness of $[0,1]$ to establish the completeness of $[0,1]$ in the above argument.

Theorem 3.6 If $A \subseteq B$ in a metric space and A is closed and B is complete, then A itself is complete.

Proof Let $a_1, a_2, a_3, \ldots$ be in A (and hence in B) with the property that $d(a_m, a_n) \to 0$ as $m, n \to \infty$. Then, by the completeness of B, $a_1, a_2, a_3, \ldots \to b$ for some $b \in B$. But as $a_1, a_2, a_3, \ldots$ are all contained in the closed set A, it also contains the limit, and so it is complete as required. □

So all closed subsets of $\mathbb{R}$ are complete. Are there any others? The set $]1,2[$ is not complete because the sequence $a_1 = 1\frac{1}{2}$, $a_2 = 1\frac{1}{4}$, $a_3 = 1\frac{1}{8}$, $\ldots$ certainly lies in the set $]1,2[$, has the property that $|a_m - a_n| \to 0$ as $m, n \to \infty$ and yet is not convergent to a point of the set. Again this is a particular case of a general result.

Theorem 3.7 Let A be a complete set in a metric space. Then A is closed.

Proof We will assume that A is not closed and show that it is not complete. So let $a_1, a_2, a_3, \ldots$ be a convergent sequence in A with its limit, a, not in A. Then, by Theorem 3.4, $d(a_m, a_n) \to 0$ as $m, n \to \infty$. But $a_1, a_2, a_3, \ldots$ does not converge to a point of A and so A is not complete. □

The results above show therefore that a subset of $\mathbb{R}$ is complete if and only if it is closed. In general, if X itself is complete (in which case (X, d) is called a *complete space*) then the concepts of closedness and completeness coincide in the metric space (X, d).

Complete sets

Exercise 36 Let $P \subseteq C(0, 1)$ be the set of all polynomials; i.e. functions of the form $x(t) = c_0 + c_1 t + c_2 t^2 + \cdots + c_n t^n$ for some real constants $c_0, c_1, c_2, \ldots, c_n$. Show that P is not closed and hence not complete in $C(0, 1)$.

The fact that $\mathbb{R}$ is complete enables us to deduce that $\mathbb{R}^2$, $\mathbb{C}$ and $\mathbb{R}^n$ are all complete.

Theorem 3.8 $\mathbb{R}^2$ is complete.

Proof Let $(x_1, y_1), (x_2, y_2), (x_3, y_3), \ldots$ be such that
$$d((x_m, y_m), (x_n, y_n)) \to 0 \text{ as } m, n \to \infty.$$
But then, since
$$0 \leqslant |x_m - x_n| = [(x_m - x_n)^2]^{1/2} \leqslant [(x_m - x_n)^2 + (y_m - y_n)^2]^{1/2}$$
$$= d((x_m, y_m), (x_n, y_n)) \to 0 \text{ as } m, n \to \infty,$$
it follows that $|x_m - x_n| \to 0$ as $m, n \to \infty$. So, by the completeness of $\mathbb{R}$, the sequence $x_1, x_2, x_3, \ldots$ converges to x, say. Similarly $y_1, y_2, y_3, \ldots \to y$, say. Then by Theorem 2.1
$$(x_1, y_1), (x_2, y_2), (x_3, y_3), \ldots \to (x, y)$$
and the completeness of $\mathbb{R}^2$ is established. □

Exercise 37 Show that $\mathbb{C}$ is complete. (You may recall from Chapter 2 that the distance from $x + yi$ to $x' + yi'$ in $\mathbb{C}$ is the same as the distance from (x, y) to (x', y') in $\mathbb{R}^2$.)

Exercise 38 Use a similar procedure to that in the proof of Theorem 3.8 to show that $\mathbb{R}^n$ is complete.

Exercise 39 Use Exercise 21 to show that every subset of a discrete space (X, d) is closed. Show also that X is itself complete and hence that every subset of X is complete.

Exercise 40 Let d, d' be metrics on a set X for which there exist positive constants α and β with
$$\alpha d'(x, y) \leqslant d(x, y) \leqslant \beta d'(x, y)$$
for all $x, y \in X$. Show that
(i) for a sequence $x_1, x_2, x_3, \ldots$ in X, $d(x_n, x_m) \to 0$ as $m, n \to \infty$ if and only if $d'(x_n, x_m) \to 0$ as $m, n \to \infty$;
(ii) for $x, x_1, x_2, x_3, \ldots$ in X, $x_1, x_2, x_3, \ldots \to x$ in (X, d) if and only if $x_1, x_2, x_3, \ldots \to x$ in (X, d');
(iii) X is complete with respect to d if and only if it is complete with respect to d'.

In Exercise 13 we introduced the metric defined on $\mathbb{R}^n$ by
$$d'(\mathbf{x},\mathbf{y}) = d'((x_1,\ldots,x_n),(y_1,\ldots,y_n)) = \max\{|x_1-y_1|,\ldots,|x_n-y_n|\}.$$
This metric space will feature in one of our applications, and again we shall need the fact that it is complete.

> **Exercise 41** Let d' be the 'max' metric on $\mathbb{R}^n$ recalled above and let d be the usual metric on $\mathbb{R}^n$. Show that
> $$d'(\mathbf{x},\mathbf{y}) \leq d(\mathbf{x},\mathbf{y}) \leq n^{1/2} d'(\mathbf{x},\mathbf{y})$$
> for all $\mathbf{x},\mathbf{y} \in X$. Deduce from Exercise 40 that $\mathbb{R}^n$ is complete with respect to the metric d'.

Our principal applications will be in the function spaces $C(a,b)$ and it will be crucial to us that these spaces are complete. We now give an outline proof of that fact.

Theorem 3.9 $C(a,b)$ is complete.

Proof Let $x_1, x_2, x_3, \ldots$ in $C(a,b)$ be such that $d(x_m, x_n) \to 0$ as $m, n \to \infty$. Then for each $t \in [a,b]$
$$0 \leq |x_m(t) - x_n(t)| \leq d(x_m, x_n) \to 0 \quad \text{as } m, n \to \infty$$
and so by Theorem 3.5 the sequence $x_1(t), x_2(t), x_3(t), \ldots$ converges, to $x(t)$, say. Doing this for each $t \in [a,b]$ defines a function $x: [a,b] \to \mathbb{R}$. We need to show that $x \in C(a,b)$ and that $d(x_n, x) \to 0$ as $n \to \infty$.

To show that $x \in C(a,b)$ we have to show that it is continuous. So let $t_1, t_2, t_3, \ldots \to t$ in $[a,b]$: we will try to deduce that $x(t_1), x(t_2), x(t_3), \ldots \to x(t)$. The continuity of each of the $x_1, x_2, x_3, \ldots$ enables us to draw the following chart of limits.

$$\begin{array}{cccccccc}
x_1(t_1) & x_1(t_2) & x_1(t_3) & \ldots & x_1(t_n) & \ldots & \to & x_1(t) \\
x_2(t_1) & x_2(t_2) & x_2(t_3) & \ldots & x_2(t_n) & \ldots & \to & x_2(t) \\
x_3(t_1) & x_3(t_2) & x_3(t_3) & \ldots & x_3(t_n) & \ldots & \to & x_3(t) \\
\vdots & \vdots & \vdots & & \vdots & & & \vdots \\
x_m(t_1) & x_m(t_2) & x_m(t_3) & \ldots & x_m(t_n) & \ldots & \to & x_m(t) \\
\vdots & \vdots & \vdots & & \vdots & & & \vdots \\
\downarrow & \downarrow & \downarrow & & \downarrow & & & \downarrow \\
x(t_1) & x(t_2) & x(t_3) & \ldots & x(t_n) & \ldots & \overset{?}{\to} & x(t).
\end{array}$$

Since this is only meant to be an outline proof we hope the reader will find it plausible from that chart that $x(t_1), x(t_2), x(t_3), \ldots \to x(t)$. For if we choose m and n large enough, then

Complete sets

$$x(t_n) \approx x_m(t_n) \approx x_m(t) \approx x(t).$$

(The analytical reader will, rightly, want a better verification that x is continuous. So, for readers with an understanding of ε-methods, here is a formal verification.

Let $\varepsilon > 0$ and let N be such that $d(x_N, x) < \varepsilon/3$. Now $x_N \in C(a,b)$ and so it is continuous. Hence, given $t_0 \in [a,b]$ there exists a $\delta > 0$ such that

$$|t - t_0| < \delta \quad \text{implies} \quad |x_N(t) - x_N(t_0)| < \varepsilon/3.$$

Therefore

$$|t - t_0| < \delta \quad \text{implies}$$

$$|x(t) - x(t_0)| \leq |x(t) - x_N(t)| + |x_N(t) - x_N(t_0)| + |x_N(t_0) - x(t_0)|$$

$$\leq d(x, x_N) + \frac{\varepsilon}{3} + d(x_N, x)$$

$$\leq \frac{\varepsilon}{3} + \frac{\varepsilon}{3} + \frac{\varepsilon}{3} = \varepsilon.$$

This formal argument involving εs and δs shows that x is continuous.)

So, having convinced ourselves (one way or another) that $x \in C(a,b)$, we must check now that x is the limit of $x_1, x_2, x_3, \ldots$ by showing that $d(x_n, x) \to 0$ as $n \to \infty$. We know that $d(x_m, x_n) \to 0$ as $m, n \to \infty$. So for $m, n \geq N_1$, say, $d(x_m, x_n) \leq 1$. Hence for any fixed $t \in [a,b]$ and $n \geq N_1$ we have

$$|x_m(t) - x_n(t)| \leq 1 \quad \text{for } m \geq N_1$$

and so

$$x_n(t) - 1 \leq x_m(t) \leq x_n(t) + 1 \quad \text{for } m \geq N_1.$$

It follows that the sequence $x_{N_1}(t), x_{N_1+1}(t), x_{N_1+2}(t), \ldots$ lies in the closed interval $[x_n(t) - 1, x_n(t) + 1]$, and hence so does the limit of the sequence, $x(t)$. Therefore

$$x_n(t) - 1 \leq x(t) \leq x_n(t) + 1$$

and

$$|x(t) - x_n(t)| \leq 1.$$

This was true for any $t \in [a,b]$ and any $n \geq N_1$. Hence $d(x_n, x) \leq 1$ for all $n \geq N_1$. A very similar argument will show that there exists an N_2 with $d(x_n, x) \leq \frac{1}{2}$ for all $n \geq N_2$; and an N_3 with $d(x_n, x) \leq \frac{1}{3}$ for all $n \geq N_3$. Continuing in this way shows that $d(x_n, x) \to 0$ as $n \to \infty$ and that $x_1, x_2, x_3, \ldots \to x$ as required.

44 The three Cs

We have thus shown that if a sequence $x_1, x_2, x_3, \ldots$ in $C(a, b)$ has the property that $d(x_m, x_n) \to 0$ as $m, n \to \infty$, then the sequence converges in $C(a, b)$. So $C(a, b)$ is complete (and so is this proof). □

In Chapter 2 we met an alternative metric on $C(a, b)$ given by

$$d(x, y) = \int_a^b |x(t) - y(t)|\, dt.$$

In this case $d(x, y)$ is the area sandwiched between the graphs of the functions x and y, as illustrated in Figure 3.3. In the final exercise of this section we see that $C(-1, 1)$ with this integral metric is not complete.

Fig. 3.3

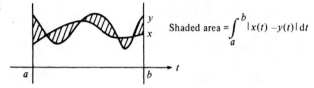

Shaded area $= \int_a^b |x(t) - y(t)|\, dt$

Exercise 42

(i) Let $x_1, x_2, x_3, \ldots$ in $C(-1, 1)$ be given by

$$x_n(t) = \begin{cases} 0 & \text{if } -1 \leqslant t \leqslant 0 \\ nt & \text{if } 0 < t < \dfrac{1}{n} \\ 1 & \text{if } \dfrac{1}{n} \leqslant t \leqslant 1, \end{cases}$$

as illustrated in Figure 3.4. Draw the graph of x_m, for some $m > n$, on the same axes and hence find $d(x_m, x_n)$, where d is the integral metric. Deduce that $d(x_m, x_n) \to 0$ as $m, n \to \infty$.

Fig. 3.4

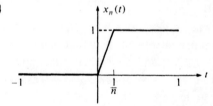

Compact sets

(ii) Now let $y: [-1, 1] \to \mathbb{R}$ be given by
$$y(t) = \begin{cases} 0 & \text{if } -1 \leq t \leq 0 \\ 1 & \text{if } 0 < t \leq 1. \end{cases}$$
Find $\int_{-1}^{1} |x_n(t) - y(t)| \, dt$ and confirm that it tends to zero as n gets large.

(iii) Assume that $x_1, x_2, x_3, \ldots \to x$ in $C(-1, 1)$ with the integral metric. Use the fact that
$$\int_{-1}^{1} |x(t) - y(t)| \, dt \leq \int_{-1}^{1} |x(t) - x_n(t)| \, dt + \int_{-1}^{1} |x_n(t) - y(t)| \, dt$$
to show that $\int_{-1}^{1} |x(t) - y(t)| \, dt = 0$. Convince yourself with a picture that it is impossible to find a *continuous* function $x: [-1, 1] \to \mathbb{R}$ such that the area sandwiched between the graphs of x and y is zero. Deduce that $x_1, x_2, x_3, \ldots$ is not convergent in $C(-1, 1)$ with the integral metric and that $C(-1, 1)$ is not complete with respect to this metric.

Before moving on to the third and last C, namely compactness, let us briefly summarise what we have learned about completeness. A complete set is one in which if a sequence has the property that the distances between terms tend to zero, then the sequence converges to a point of the set. A complete set is necessarily closed, but a closed set need not be complete. But if we restrict attention to complete metric spaces (X, d); i.e. where X itself is complete, then subsets of X are complete if and only if they are closed. All the principal spaces which we meet in our applications ($\mathbb{R}$, $\mathbb{R}^2$, $\mathbb{R}^n$, $\mathbb{C}$ and $C(a, b)$ with their usual metrics) are complete and so all their closed subsets are complete.

3.5 Compact sets

The third property of sets which we wish to consider is that of 'compactness'. This is rather less relevant to us than the previous properties of being closed or complete, so the reader may turn straight to Chapter 4 without much loss. However, having come so close to this property we can consider it with little extra effort. Then our applications in the next chapter will have several extensions and in the final chapter the relevance of compactness to real analysis will become clear.

We remarked in the previous section that any real sequence in the interval $[a, b]$ has a convergent subsequence (with its limit in the set). Indeed, it was that property of $[a, b]$ which enabled us to show that $\mathbb{R}$ is complete. This property has such far-reaching consequences in analysis that we isolate it in the following definition.

Definition A set A in a metric space (X, d) is *compact* if every sequence in A has a subsequence convergent to a point of A.

The closed bounded intervals $[a,b]$ in $\mathbb{R}$ are therefore compact, but $\mathbb{R}$ itself is not. For example, the sequence 1, 2, 3, 4, ... has no convergent subsequence. In fact we shall see in Theorem 3.11 that the compact subsets of $\mathbb{R}$ are precisely the closed and bounded sets. But first we provide the missing link in the implications

$$\text{compact} \Rightarrow \text{complete} \Rightarrow \text{closed}$$

which shows that our three properties are progressively stronger.

Theorem 3.10 Let A be a compact set in a metric space. Then A is complete.

Proof Let A be compact in (X, d) and let $a_1, a_2, a_3, \ldots$ be a sequence in A with the property that $d(a_m, a_n) \to 0$ as $m, n \to \infty$. To establish the completeness of A we must show that this sequence converges to a point of A. By compactness there is a convergent subsequence

$$a_{k_1}, a_{k_2}, a_{k_3}, \ldots \to a$$

with its limit in A. It is now easy to show that the original sequence also converges to a. For

$$0 \leq d(a_n, a) \leq d(a_n, a_{k_n}) + d(a_{k_n}, a) \to 0$$
$$\downarrow \qquad\qquad \downarrow$$
$$0 \qquad\qquad 0$$

since $d(a_n, a_m) \to 0$ as $m, n \to \infty$; since $a_{k_1}, a_{k_2}, a_{k_3}, \ldots$ converges to a

and so $d(a_n, a) \to 0$, i.e. $a_1, a_2, a_3, \ldots$ itself converges to the point a of A. Hence A is complete. □

We saw in Theorem 3.6 that a closed subset of a complete set is complete. You can use a similar approach in the following exercise.

Exercise 43 Let $A \subseteq B$ in a metric space, where A is closed and B is compact. Show that A is compact.

Exercise 44 Let d be the discrete metric on a set X.

Remembering that a sequence converges in this space if and only if it is of the form

$$x_1, x_2, \ldots, x_N, x, x, x, \ldots,$$

show that $A \subseteq X$ is compact if and only if A contains only a finite number of points.

Exercise 45 Which of the following subsets of $\mathbb{R}$ are compact?
(a) $\mathbb{Q} \cap [0,1]$ (i.e. the fractions in $[0,1]$)
(b) $[2, 2\frac{1}{2}] \cup [3, 3\frac{1}{3}] \cup [4, 4\frac{1}{4}] \cup \ldots$
(c) $\{1, 2, 3, \ldots, N\}$.

In Exercise 45 the set in (a) is not closed (for example, it has a sequence converging to $1/\sqrt{2}$ which is outside the set), and the set in (b) contains the sequence $2, 3, 4, \ldots$ which has no convergent subsequence. So only the set in (c) is compact and, as you may have realised from your solutions to Exercises 44 and 45, it is easy to check that any set containing only a finite number of points in a metric space is compact.

We are now able to prove that, as far as $\mathbb{R}$ is concerned, the compact sets are just the closed and bounded ones.

Theorem 3.11 Let $A \subseteq \mathbb{R}$. Then A is compact if and only if it is closed and bounded.

Proof First let $A \subseteq \mathbb{R}$ be compact. Then, as we saw in Theorems 3.10 and 3.7,

compact $\Rightarrow$ complete $\Rightarrow$ closed,

and so A is certainly closed. Also A can have no sequence of points tending to infinity (or $-\infty$) since such sequences do *not* have any convergent subsequences. Hence A is bounded.

Conversely, let $A \subseteq \mathbb{R}$ be closed and bounded. Then A is a subset of some interval $[a, b]$. Hence A is a closed subset of a compact set and (by Exercise 43) is itself compact. □

Exercise 46 Let $A \subseteq \mathbb{R}$ be non-empty and compact. Show that A has a least member and a greatest member. (You know that A is bounded, so it has a least upper bound, α, say. There must be a sequence of members of A converging to α.)

Although in $\mathbb{R}$ compactness is characterised so simply, it is not the

48 The three Cs

case in all spaces that closed and bounded sets are necessarily compact. But of course to consider this we need to know what is meant in general by a 'bounded' set. As its name implies, a bounded set is one where the distance between any pair of its points has certain bounds.

Definition A set A in a metric space (X, d) is *bounded* if there exists a number D such that $d(a, a') \leqslant D$ for all $a, a' \in A$.

Exercise 47 Show that a set $A \subseteq \mathbb{R}^2$ is bounded if and only if there exist numbers a, b, c and d such that $(x, y) \in A$ implies $x \in [a, b]$ and $y \in [c, d]$. (Figures 3.5 and 3.6 might help.)

Exercise 48 Show that a non-empty set A in a metric space (X, d) is bounded if and only if there exists $a \in A$ and a number D' such that $d(a, a') \leqslant D'$ for all $a' \in A$. (This is like the definition but with a fixed.)

Exercise 49 Show that if the points $a, a_1, a_2, a_3, \ldots$ of a metric space (X, d) have the properties

Fig. 3.5

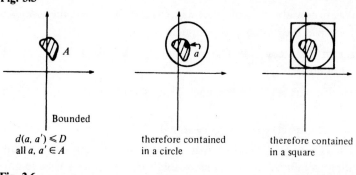

$d(a, a') \leqslant D$
all $a, a' \in A$

Bounded

therefore contained in a circle

therefore contained in a square

Fig. 3.6

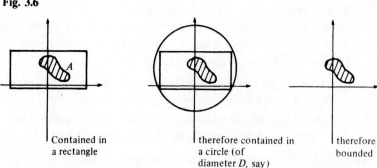

Contained in a rectangle

therefore contained in a circle (of diameter D, say)

therefore bounded

Compact sets

$d(a, a_1) \geq 1$, $d(a, a_2) \geq 2$, $d(a, a_3) \geq 3$, ...,

then the sequence a_1, a_2, a_3, ... has no convergent subsequence. Deduce that in a metric space a compact set is bounded.

So, in any metric space, a compact set is both closed and bounded; and, in $\mathbb{R}$ at least, the converse is true. This simple characterisation also holds in $\mathbb{R}^2$.

Theorem 3.12 Let $A \subseteq \mathbb{R}^2$. Then A is compact if and only if it is closed and bounded.

Proof If A is compact, then (as we have seen in general) it must be closed and bounded.

Conversely, let A be closed and bounded. Then as we saw in Exercise 47 there exist numbers a, b, c and d such that A is a closed subset of the set

$$B = \{(x, y): x \in [a, b] \text{ and } y \in [c, d]\}.$$

So if we can show that B is compact, then it will follow that A is a closed subset of a compact set and hence is itself compact.

So let (x_1, y_1), (x_2, y_2), (x_3, y_3), ... be a sequence of points in B. Then $x_1, x_2, x_3, \ldots$ is a sequence of real numbers in $[a, b]$ and, by the compactness of that set, there is a subsequence $x_{k_1}, x_{k_2}, x_{k_3}, \ldots$ convergent to a point x of $[a, b]$. Then the sequence $y_{k_1}, y_{k_2}, y_{k_3}, \ldots$ of real numbers in the compact set $[c, d]$ must have a convergent subsequence $y_{l_1}, y_{l_2}, y_{l_3}, \ldots$ with its limit y in $[c, d]$. But then

$$x_{l_1}, x_{l_2}, x_{l_3}, \ldots \to x \in [a, b],$$
$$y_{l_1}, y_{l_2}, y_{l_3}, \ldots \to y \in [c, d],$$

and so

$$(x_{l_1}, y_{l_1}), (x_{l_2}, y_{l_2}), (x_{l_3}, y_{l_3}), \ldots \to (x, y) \in B.$$

Hence any sequence in B has a convergent subsequence (with limit in B) and so B is compact. Finally, as remarked above, it follows that A is compact because it is a closed subset of B. $\square$

Exercise 50 Show that $A \subseteq \mathbb{R}^3$ is bounded if and only if there exist real numbers a_1, a_2, a_3, b_1, b_2 and b_3 with

$$A \subseteq \{(x, y, z) \in \mathbb{R}^3: x \in [a_1, b_1], y \in [a_2, b_2] \text{ and } z \in [a_3, b_3]\}.$$

Hence show that A is compact if and only if it is closed and bounded. (The keen reader could confirm that the same is true for $\mathbb{R}^n$.)

So we have seen that in $\mathbb{R}^n$ compactness is easily characterised. We have also seen that, in any metric space, a compact set must be closed and bounded. But is it true in general that a closed and bounded set must be compact?

> **Exercise 51** Let A be a set in a metric space (X, d) and assume that A contains a sequence $a_1, a_2, a_3, \ldots$ such that, for some positive number δ, $d(a_m, a_n) \geqslant \delta$ for each $m \neq n$. Show that A is not compact.
>
> **Exercise 52** Let $A \subseteq C(0,1)$ consist of those continuous functions $[0,1] \to [0,1]$. Show that any two members of A are at most a distance 1 apart. Deduce (using Exercise 28) that A is closed and bounded. By considering functions like
>
> $$a_n(t) = \begin{cases} 2^n t & 0 \leqslant t \leqslant \frac{1}{2^n} \\ 1 & \frac{1}{2^n} < t \leqslant 1 \end{cases}$$
>
> show that A is not, however, compact.

It seems that when we move to spaces other than $\mathbb{R}^n$ compactness is less straightforward. For example the set A in Exercise 52 is closed and bounded, and yet it fails to be compact. We confirmed this by finding an infinite number of members of A each mutually at least distance $\frac{1}{2}$ apart.

In the final chapter the relevance of compactness to continuity will become clear. But at this stage we are able to see a preview of one of those results in some exercises (the second of which is actually needed in the next chapter).

> **Exercise 53** Let A be a compact subset of $\mathbb{R}$ and let $f: A \to \mathbb{R}$ be continuous. Show that for any $a_1, a_2, a_3, \ldots$ in A the real sequence $f(a_1), f(a_2), f(a_3), \ldots$ has a convergent subsequence with limit $f(a_0)$ for some $a_0 \in A$. Deduce that the set $f(A) = \{f(a): a \in A\}$ is compact.
>
> **Exercise 54** Let A be a non-empty compact subset of the metric space (X, d), let L be any fixed real number, and let $F: A \to \mathbb{R}$ be a function with the property that
> $$|F(x) - F(y)| \leqslant L d(x, y)$$

Compact sets

for all $x, y \in A$. Show that for any sequence $a_1, a_2, a_3, \ldots$ in A the real sequence $F(a_1), F(a_2), F(a_3), \ldots$ has a convergent subsequence with limit $F(a_0)$ for some $a_0 \in A$. Deduce that the set $F(A) = \{F(a): a \in A\}$ is compact in $\mathbb{R}$. Hence show that there exists $a_0 \in A$ such that $F(a_0)$ is less than or equal to every other $F(a)$ for $a \in A$.

Exercise 55 Let A be a compact subset of the metric space (X, d), let (X', d') be another metric space, let L be any fixed real number, and let $f: A \to X'$ be a function with the property that

$$d'(f(x), f(y)) \leqslant L d(x, y)$$

for all $x, y \in A$. Show that $f(A)$ is compact in (X', d').

This chapter has been, of necessity, full of the 'bread and butter' results of the subject. But your patience in getting this far will be rewarded because we are now able to apply our newly found techniques to a wide range of problems.

4

The contraction mapping principle

4.1 Real fixed points

Consider the graph of a continuous real function $f: [a, b] \to [a, b]$. Note that, for these purposes, 'continuous' simply means that the graph is in one piece, and the fact that $f: [a, b] \to [a, b]$ means that it is defined for all x with $a \leq x \leq b$ and that $f(x)$ then satisfies $a \leq f(x) \leq b$ (see Figure 4.1). Consider, too, the line $y = x$ (dotted in the picture). When $x = a$ the graph of f is on or above the line (since $f(a)$, being in $[a, b]$, is at least a). When $x = b$ the graph of f is on or below the line (since $f(b)$, being in $[a, b]$, is at most b). Since the graph of f is in one piece, it means that at some point in $[a, b]$ the graph of f crosses the line; i.e. for some $x \in [a, b]$ we have $x = f(x)$. (To those readers who remember their basic real analysis: this is, in fact, based on the 'intermediate value theorem'.)

Exercise 56 Let $f: [-\pi/2, \pi/2] \to \mathbb{R}$ be given by $f(x) = \cos 4x$. Sketch the graph of f and observe that $f(x) \in [-\pi/2, \pi/2]$ for each x (which is not to say, of course, that every number in $[-\pi/2, \pi/2]$ occurs as an $f(x)$). In how many places does the graph of f cross the line $y = x$? How

Fig. 4.1

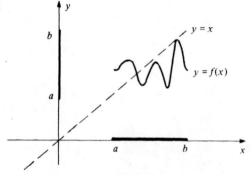

Real fixed points

many roots in $[-\pi/2, \pi/2]$ does the equation $x = \cos 4x$ have?

So the graph of any continuous function $f: [a, b] \to [a, b]$ crosses the line $y = x$ in at least one place; in other words, there is at least one $x \in [a, b]$ for which $x = f(x)$. So for any continuous function $f: [a, b] \to [a, b]$ the equation $x = f(x)$ has at least one root. But is that pure mathematical fact of any use in actually calculating the roots?

Exercise 57 Let $f: \mathbb{R} \to \mathbb{R}$ be given by $f(x) = \cos x$. Sketch the graph of f and observe that it crosses the line $y = x$ in just one place. Choose any real number x_1 and (with your calculator set in radians) work out $\cos x_1$, $\cos(\cos x_1)$, $\cos(\cos(\cos x_1))$, ... ; i.e. display your number and repeatedly press the cos button. Continue until consecutive answers agree to five decimal places.

What you have found in Exercise 57 (to five decimal places, anyway) is a number unaffected by taking its cosine; i.e. an approximate root of the equation $\cos x = x$. Check for yourselves that $\cos 0.73908 \approx 0.73908$. Figure 4.2 shows the position of $x_2 = \cos x_1$ in relation to x_1. So if we start with any x_1 and keep reapplying the cosine function, then the sequence

$$x_1, \quad x_2 = \cos x_1, \quad x_3 = \cos x_2, \quad x_4 = \cos x_3, \quad \ldots$$

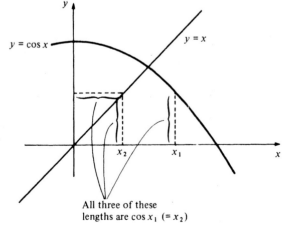

Fig. 4.2

All three of these lengths are $\cos x_1$ $(= x_2)$

Fig. 4.3

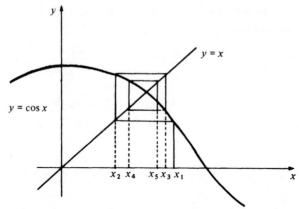

can be illustrated as in Figure 4.3. And, as we saw with our earlier calculations, the sequence does seem to converge to that unique x for which $\cos x = x$.

This is precisely the sort of sequence with which we started the book. We are now ready to ask 'what property of f will ensure that a solution to $x = f(x)$ can be found by starting with any number x_1 and repeatedly applying the function f to give

$$x_1, \quad x_2 = f(x_1), \quad x_3 = f(x_2), \quad x_4 = f(x_3), \quad \ldots$$

leading to a root?'

For what function f can we be sure that if x_1 is a guess at a root of $x = f(x)$, then $x_2 = f(x_1)$ is closer to the root? Suppose that x_0 is the root of $x = f(x)$: then we require that

$$|x_2 - x_0| < |x_1 - x_0| \quad \text{or} \quad |f(x_1) - f(x_0)| < |x_1 - x_0|.$$

Since we do not know the value of x_0 we shall have to require that

$$|f(x) - f(y)| < |x - y|$$

for all $x \neq y$; i.e. f reduces the distance between points.

The cosine function certainly has this property. For any distinct x and y

$$\begin{aligned}
|\cos x - \cos y| &= \left| 2 \sin\left(\frac{x-y}{2}\right) \sin\left(\frac{x+y}{2}\right) \right| \\
&= 2 \left|\sin\left(\frac{x-y}{2}\right)\right| \left|\sin\left(\frac{x+y}{2}\right)\right| \\
&\leq 2 \left|\sin\left(\frac{x-y}{2}\right)\right|
\end{aligned}$$

$$< 2\left|\frac{x-y}{2}\right| \quad \text{(since } |\sin w| < |w| \text{ for all } w \neq 0)$$
$$= |x-y|.$$

Exercise 58 Let $f: \mathbb{R} \to \mathbb{R}$ be given by $f(x) = \sin x$. Show that, for any distinct x and y,
$$|f(x) - f(y)| < |x-y|.$$
Choose any number x_1 and enter it on your calculator (set in radians). Repeatedly apply the sine function and observe that the displayed numbers seem to converge to 0. This is because 0 is the unique root of $x = \sin x$.

We saw earlier that for the function given by $f(x) = \cos 4x$ the equation $x = f(x)$ has three roots. If, however, a function has the property that
$$|f(x) - f(y)| < |x-y|$$
for each $x \neq y$, then the equation $x = f(x)$ can have at most one root.

Exercise 59 Let X be a subset of $\mathbb{R}$ and let $f: X \to \mathbb{R}$ have the property that
$$|f(x) - f(y)| < |x-y|$$
for each $x \neq y$. Show that if $x_0 = f(x_0)$ and $x'_0 = f(x'_0)$ then $x_0 = x'_0$. Deduce that the equation $x = f(x)$ has at most one root.

We mention in passing that it is possible, however, for a function to have the property that
$$|f(x) - f(y)| < |x-y|$$
for each $x \neq y$ and yet for the equation $x = f(x)$ to have no solution.

Exercise 60 Let $f: [1, \infty[\to [1, \infty[$ be given by $f(x) = x + 1/x$. Show that
$$|f(x) - f(y)| < |x-y|$$
for each $x \neq y$, but that there is no x for which $x = f(x)$.

In Exercise 60, for very large x and y, $|f(x) - f(y)|$ gets very close indeed to $|x-y|$. If we want our equations $x = f(x)$ to have a root perhaps we ought to ensure that $|f(x) - f(y)|$ cannot get close to $|x-y|$.

Exercise 61 Let $f: [1, \infty[\to [1, \infty[$ be given by $f(x) = \frac{25}{26}(x + 1/x)$. Show that
$$|f(x) - f(y)| \leq \tfrac{25}{26}|x - y|$$
for all $x, y \in [1, \infty[$. Display any number x_1 in your calculator and work out
$$x_1, \quad x_2 = f(x_1), \quad x_3 = f(x_2), \quad \ldots$$
Observe that your sequence seems to be converging to 5. Confirm, by solving the equation algebraically, that 5 is the unique root of the equation $x = f(x)$.

So although the condition $|f(x) - f(y)| < |x - y|$ was not enough to ensure a root of the equation $x = f(x)$, perhaps the condition
$$|f(x) - f(y)| \leq k|x - y|$$
(where k is some number less than 1) *will* be enough.

Exercise 62 Let $f: [0, 1] \to \mathbb{R}$ be given by $f(x) = \tfrac{1}{7}(x^3 + x^2 + 1)$. Show that $f(x)$ is actually in $[0, 1]$ and hence that $f: [0, 1] \to [0, 1]$. Show also that for any $x, y \in [0, 1]$
$$|f(x) - f(y)| \leq \tfrac{5}{7}|x - y|.$$
(You will need the fact that
$$|x^3 - y^3| = |x - y| \, |x^2 + xy + y^2| \text{ etc.})$$
Calculate $f(0), f(f(0)), f(f(f(0))), \ldots$ and hence find, to three decimal places, a root of the equation $x = f(x)$. Show that this is an approximate root of
$$x^3 + x^2 - 7x + 1 = 0.$$

We have used informal arguments in this section in order to prepare ourselves for the formality of the next section. But in our informal way we have seen that a condition like
$$|f(x) - f(y)| \leq k|x - y|$$
(for some number $k < 1$) might ensure that the equation $x = f(x)$ has a root. Then, starting with any number x_1, the iterative process of repeatedly applying f to x_1 might lead us to that root.

The points for which $x = f(x)$ are called the *fixed points* of f and we shall now see how, in some very general situations, if f reduces distances (rather like we have seen) then f has a unique fixed point. We shall also see how iterating with f leads us to that fixed point.

4.2 Contractions

Definition Let (X, d) be a metric space. Then a *contraction* of (X, d) is a function $f: X \to X$ with the property that, for some real number $k < 1$,
$$d(f(x), f(y)) \leq k d(x, y) \quad \text{for all } x, y \in X.$$

So, as we saw in Exercise 61, the function $f: [1, \infty[\to [1, \infty[$ given by $f(x) = \frac{25}{26}(x + 1/x)$ is a contraction since (as we checked earlier)
$$d(f(x), f(y)) = |f(x) - f(y)| \leq \tfrac{25}{26}|x - y| = \tfrac{25}{26} d(x, y).$$
So, with the usual metric and with $k = \frac{25}{26}$, this function is a contraction of $[1, \infty[$. Similarly, as we saw in Exercise 62, the function $f: [0, 1] \to [0, 1]$ given by $f(x) = \frac{1}{7}(x^3 + x^2 + 1)$ is a contraction since
$$d(f(x), f(y)) \leq \tfrac{5}{7} d(x, y).$$

In both cases those contractions had unique fixed points, found by taking any initial guess and reapplying the function. But, of course, the advantage of our general definition is that it extends to worlds beyond $\mathbb{R}$.

Exercise 63 Use the properties
$$|\cos x - \cos y| \leq |x - y| \quad \text{and} \quad |\sin x - \sin y| \leq |x - y|$$
derived earlier to show that
$$\left[\left(\frac{\cos y_2 - \cos x_2}{2}\right)^2 + \left(\frac{\sin y_1 - \sin x_1}{2}\right)^2\right]^{1/2}$$
$$\leq \tfrac{1}{2}[(y_1 - x_1)^2 + (y_2 - x_2)^2]^{1/2}.$$
Deduce that if $f: \mathbb{R}^2 \to \mathbb{R}^2$ is given by
$$f(\mathbf{x}) = f(x_1, x_2) = (\tfrac{1}{2} \cos x_2, \tfrac{1}{2} \sin x_1 + 1)$$
(and d is the usual metric on $\mathbb{R}^2$), then
$$d(f(\mathbf{x}), f(\mathbf{y})) \leq \tfrac{1}{2} d(\mathbf{x}, \mathbf{y})$$
for each $\mathbf{x} = (x_1, x_2)$ and $\mathbf{y} = (y_1, y_2)$ in $\mathbb{R}^2$. This shows that f is a contraction of $\mathbb{R}^2$.

We saw with real functions that iterating with a contraction f led to the unique fixed point of f.

Exercise 64 Let $f: \mathbb{R}^2 \to \mathbb{R}^2$ be given by
$$f(x) = f(x_1, x_2) = (\tfrac{1}{2} \cos x_2, \tfrac{1}{2} \sin x_1 + 1),$$
as in the previous exercise. Let $\mathbf{x}_1 = (0, 0)$. Then
$$f(\mathbf{x}_1) = f(0, 0) = (\tfrac{1}{2} \cos 0, \tfrac{1}{2} \sin 0 + 1) = (0.5, 1).$$

Let $\mathbf{x}_2 = f(\mathbf{x}_1) = (0.5, 1)$. Continue in this way (so that, for example,
$$\mathbf{x}_3 = f(\mathbf{x}_2) = f(0.5, 1) = (\tfrac{1}{2}\cos 1, \tfrac{1}{2}\sin 0.5 + 1) = \ldots)$$
until two consecutive pairs agree to two decimal places. What you hope to have found, approximately, is a root of $\mathbf{x} = f(\mathbf{x})$, i.e.
$$(x_1, x_2) = (\tfrac{1}{2}\cos x_2, \tfrac{1}{2}\sin x_1 + 1).$$
Confirm with your calculator that the pair (x_1, x_2) which (to two decimal places) repeated itself in the above process does give
$$x_1 \approx \tfrac{1}{2}\cos x_2 \quad \text{and} \quad x_2 \approx \tfrac{1}{2}\sin x_1 + 1.$$

Already we have seen that by having a contraction of $\mathbb{R}^2$ rather than $\mathbb{R}$ we can extend the sort of equations to which our methods will apply.

Now we are going to ask whether a contraction of any metric space will have a unique fixed point and, if so, whether the sequence
$$x_1, \quad x_2 = f(x_1), \quad x_3 = f(x_2), \quad \ldots$$
leads to that fixed point. We saw earlier that a sequence which appears to be behaving nicely might not actually converge because the metric space might have some gaps.

Exercise 65 Let $X = \{x \in \mathbb{Q} : x \geq 1\}$, the set of rational numbers of 1 or more, and define $f: X \to X$ by $f(x) = x/2 + 1/x$. Show that, for all $x, y \in X$,
$$d(f(x), f(y)) \leq \tfrac{1}{2}d(x, y)$$
(where d is the usual metric for real numbers). Show, however, that for this contraction f there is no $x \in X$ for which $x = f(x)$.

All these clues which we have assembled together seem to imply that we need to consider contractions of a complete metric space, bringing us at last to the key result of this chapter (and the focal point of our whole approach). It is known as *the contraction mapping principle* (or *Banach's fixed-point principle*).

Theorem 4.1 Let $f: X \to X$ be a contraction of the complete metric space (X, d). Then f has a unique fixed point. Furthermore, if x_1 is any point of X, then the sequence

Contractions

$$x_1, \quad x_2 = f(x_1), \quad x_3 = f(x_2), \quad \ldots$$

converges to that unique fixed point.

Proof Let x_1 be any point of X and, as usual, let $x_2 = f(x_1), x_3 = f(x_2)$, and so on. We hope to prove that the sequence $x_1, x_2, x_3, \ldots$ converges and that its limit is the unique fixed point of f.

Given a sequence in a complete metric space the obvious way to test convergence is by seeing whether the sequence is a Cauchy sequence. Calculating the distance from x_n to x_m is a little ambitious, so let us start by considering the distance from x_n to x_{n+1}. From the inequalities

$$d(x_n, x_{n+1}) = d(f(x_{n-1}), f(x_n)) \leqslant k d(x_{n-1}, x_n)$$
$$= k d(f(x_{n-2}), f(x_{n-1})) \leqslant k^2 d(x_{n-2}, x_{n-1})$$
$$\vdots$$
$$= k^{n-2} d(f(x_1), f(x_2)) \leqslant k^{n-1} d(x_1, x_2)$$

we see that $d(x_n, x_{n+1}) \leqslant k^{n-1} d(x_1, x_2)$. (And, since $0 \leqslant k < 1$, for large n the distance between x_n and x_{n+1} is very small.)

We can now extend this easily to the distance from x_n to x_m for any $m > n$. For

$$d(x_n, x_m) \leqslant d(x_n, x_{n+1}) + d(x_{n+1}, x_m)$$
$$\leqslant d(x_n, x_{n+1}) + d(x_{n+1}, x_{n+2}) + d(x_{n+2}, x_m)$$
$$\vdots$$
$$\leqslant d(x_n, x_{n+1}) + d(x_{n+1}, x_{n+2})$$
$$+ d(x_{n+2}, x_{n+3}) + \cdots + d(x_{m-1}, x_m).$$

And we have already established an upper bound for the distance between successive xs. Hence

$$d(x_n, x_m) \leqslant d(x_n, x_{n+1}) + d(x_{n+1}, x_{n+2}) + \cdots + d(x_{m-1}, x_m)$$
$$\leqslant k^{n-1} d(x_1, x_2) + k^n d(x_1, x_2) + \cdots + k^{m-2} d(x_1, x_2)$$
$$\leqslant (k^{n-1} + k^n + \ldots) d(x_1, x_2)$$
$$= k^{n-1}(1 + k + k^2 + \ldots) d(x_1, x_2)$$
$$= \frac{k^{n-1}}{1-k} d(x_1, x_2) \quad (\to 0 \text{ as } n \to \infty).$$

So, as we hoped, if $n \to \infty$ and $m > n$, then $d(x_n, x_m) \to 0$; i.e. $x_1, x_2, x_3, \ldots$ is a Cauchy sequence in the complete metric space (X, d). Hence

$$x_1, x_2, x_3, \ldots \to x \quad \text{(say)}$$

therefore

$f(x_1), f(x_2), f(x_3), \ldots \to f(x)$ since the distance between $f(x_n)$ and $f(x)$ is, if anything, less than that of x_n from x;

i.e. $x_2, x_3, x_4, \ldots \to f(x)$ since, by our original construction, $f(x_1) = x_2$ and $f(x_2) = x_3$, etc.

Also $x_2, x_3, x_4, \ldots \to x$ from the last line of the previous page

Therefore $x = f(x)$ for sequences cannot have two different limits.

So the limit of the sequence is indeed a fixed point of f. The fact that it is the unique fixed point of f (and hence that, regardless of the choice of x_1, the sequence will always lead to this fixed point) is left as an exercise:

Exercise 66 Let (X, d) be a metric space and let $f: X \to X$ have the property that
$$d(f(x), f(y)) < d(x, y)$$
for all $x \neq y$ in X. Show that if $f(x_0) = x_0$ and $f(x_0') = x_0'$, then $x_0 = x_0'$; i.e. f has at most one fixed point.

That final fact now completes the proof of Theorem 4.1. □

We shall see several applications of the contraction mapping principle in later sections, but we also include three here as exercises before proceeding to some extensions of the result.

Exercise 67 Let $f: \mathbb{R}^3 \to \mathbb{R}^3$ be given by
$$f(\mathbf{x}) = f(x_1, x_2, x_3) = (\tfrac{1}{2} \cos x_2 + 1, \tfrac{2}{3} \sin x_3, \tfrac{3}{4} x_1).$$
Show that f is a contraction of $\mathbb{R}^3$ and deduce that the simultaneous equations
$$x_1 = \tfrac{1}{2} \cos x_2 + 1,$$
$$x_2 = \tfrac{2}{3} \sin x_3,$$
$$x_3 = \tfrac{3}{4} x_1$$
have a unique solution. Find, by an iterative process, an approximation to that solution.

Contractions

Exercise 68 Recall that $C(0, \frac{1}{2})$ is the collection of continuous functions from $[0, \frac{1}{2}]$ to $\mathbb{R}$ together with the 'max' metric. So each x in $C(0, \frac{1}{2})$ is a function for which each $x(t)$ is defined $(0 \leq t \leq \frac{1}{2})$. Given such an x, define a new function $f(x)$ in $C(0, \frac{1}{2})$ by
$$(f(x))(t) = t(x(t) + 1).$$
So, for example, if $x(t) = t^2 + 1$ for $0 \leq t \leq \frac{1}{2}$, then $f(x)$ is the function given by $(f(x))(t) = t(t^2 + 2)$. Show that f is a contraction of $C(0, \frac{1}{2})$. Show also that if x_1 is the function given by $x_1(t) = t$, then iterating with f yields the sequence of functions $x_1, x_2, x_3, \ldots$ where
$$x_n(t) = t + t^2 + \cdots + t^n.$$
Use the sum of a geometric progression to show that, for each t, $x_n(t)$ tends to $t/(1-t)$ as n tends to infinity. Confirm directly that the function x given by $x(t) = t/(1-t)$ is the unique fixed point of f.

Our applications to $C(a, b)$ will, in general, be much more significant and useful than that particular exercise. We now see an application to matrices.

Exercise 69 Let $\begin{pmatrix} a & b \\ c & d \end{pmatrix}$ be a real matrix all of whose entries are, in modulus, less than $\frac{1}{2}$. Let $f: \mathbb{R}^2 \to \mathbb{R}^2$ be given by
$$f(\mathbf{x}) = f(x_1, x_2) = (x_1', x_2')$$
where
$$\begin{pmatrix} x_1' \\ x_2' \end{pmatrix} = \begin{pmatrix} a & b \\ c & d \end{pmatrix} \begin{pmatrix} x_1 \\ x_2 \end{pmatrix}.$$
Show that
$$\max\{|x_1'|, |x_2'|\} \leq 2 \max\{|a|, |b|, |c|, |d|\} \cdot \max\{|x_1|, |x_2|\}.$$
Deduce that f is a contraction of the metric space $\mathbb{R}^2$ with metric given by
$$d(\mathbf{x}, \mathbf{y}) = d((x_1, x_2), (y_1, y_2)) = \max\{|x_1 - y_1|, |x_2 - y_2|\}.$$
As we saw in Exercise 41 this space is complete and so f has a unique fixed point; i.e. there is precisely one $(x_1, x_2) \in \mathbb{R}^2$ with

62 The contraction mapping principle

$$\begin{pmatrix} x_1 \\ x_2 \end{pmatrix} = \begin{pmatrix} a & b \\ c & d \end{pmatrix} \begin{pmatrix} x_1 \\ x_2 \end{pmatrix}.$$

This (x_1, x_2) must clearly be $(0, 0)$. Deduce that

$$\begin{pmatrix} a-1 & b \\ c & d-1 \end{pmatrix}$$

is non-singular.

Exercise 70 The interested reader can extend the ideas of the previous exercise to show that if M is a real $n \times n$ matrix all of whose entries are, in modulus, less than $1/n$, then there exists precisely one $(x_1, x_2, \ldots, x_n) \in \mathbb{R}^n$ with

$$\begin{pmatrix} x_1 \\ x_2 \\ \vdots \\ x_n \end{pmatrix} = M \begin{pmatrix} x_1 \\ x_2 \\ \vdots \\ x_n \end{pmatrix},$$

and that $M - I$ is non-singular, where I is the $n \times n$ identity matrix.

Those last two exercises give the reader a taste of the applications to linear algebra.

4.3 Real contractions revisited

Most of the real functions which we have encountered have been differentiable and, for such functions, there is an easy test of whether or not they are contractions (with respect to the usual real metric).

Theorem 4.2 Let $f: [a, b] \to [a, b]$ be differentiable. Then f is a contraction of $[a, b]$ if and only if there exists a number $k < 1$ with $|f'(x)| \leq k$ for all $x \in]a, b[$.

Proof Assume first that f is a contraction, i.e. that there exists $k < 1$ with $|f(x) - f(y)| \leq k|x - y|$ for all $x, y \in [a, b]$. Then, in particular, for any x and $x + \delta x$ in $[a, b]$ we have

$$|f(x + \delta x) - f(x)| \leq k|(x + \delta x) - x| = k|\delta x|.$$

Hence, for $\delta x \neq 0$,

$$\left| \frac{f(x + \delta x) - f(x)}{\delta x} \right| \leq k$$

and the limit of the left-hand expression as $\delta x \to 0$ is also less than or

Real contractions revisited

equal to k. But that limit is precisely $|f'(x)|$. Hence the bound on $|f'(x)|$ is proved.

Conversely, assume that $|f'(x)| \leq k$ (<1) for all $x \in]a, b[$. For any $x \neq y$ in $[a, b]$ we know that

$$\frac{f(x) - f(y)}{x - y} = f'(c)$$

for some c between x and y. This is the *mean value theorem* from real analysis, but for those readers who wish to at least be convinced of its plausibility there is a diagram (Figure 4.4).

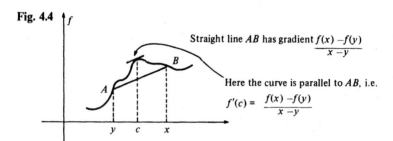

Fig. 4.4

Straight line AB has gradient $\frac{f(x) - f(y)}{x - y}$

Here the curve is parallel to AB, i.e. $f'(c) = \frac{f(x) - f(y)}{x - y}$

But we know that $|f'(c)| \leq k$ and so

$$\left|\frac{f(x) - f(y)}{x - y}\right| = f'(c) \leq k.$$

Hence
$$|f(x) - f(y)| \leq k|x - y|$$

and f is a contraction, as required. □

Exercise 71 Show that the function $f: \mathbb{R} \to \mathbb{R}$ given by $f(x) = \cos x$ is not a contraction, but that the function $g(x) = \frac{99}{100} \cos x$ is a contraction.

Exercise 72 Confirm by differentiation that the function $f: [1, \infty[\to [1, \infty[$ given by $f(x) = x + 1/x$ is not a contraction, but that the function $g: [1, \infty[\to [1, \infty[$ given by $g(x) = \frac{25}{26}(x + 1/x)$ is.

Exercise 73 Let α be any real number larger than 5. Show by differentiation that $f: [-1, 1] \to [-1, 1]$ given by

$$f(x) = \frac{1}{\alpha}(x^3 + x^2 + 1)$$

is a contraction of $[-1,1]$. Deduce that the equation
$$x^3+x^2-\alpha x+1=0$$
subject to the restriction $|x|\leqslant 1$ has exactly one root.

Exercise 74 Show that the function $f\colon \mathbb{R}\to\mathbb{R}$ given by $f(x)=\cos(\cos x)$ is a contraction, but that the function $g\colon\mathbb{R}\to\mathbb{R}$ given by
$$g(x)=\sin(\sin(\sin(\ldots(\sin x)\ldots)))$$
(with any number of sines) is not a contraction.

4.4 Some extensions

If $f\colon X\to X$, then $f\circ f$ is given by $f\circ f(x)=f(f(x))$ and it is again a function from X to X: it is called the second *iterate* of f. In general $f\circ f\circ\ldots\circ f$ (with N fs) is the Nth iterate and, of course, is given by
$$f\circ f\circ\ldots\circ f(x)=f(f(\ldots f(x)\ldots)).$$
It is abbreviated to f^N. So, for example, the third iterate of $f\colon\mathbb{R}\to\mathbb{R}$ defined by $f(x)=x^2$ is given by
$$f^3(x)=f(f(f(x)))=((x^2)^2)^2=x^8,$$
and the fourth iterate of $g\colon\mathbb{R}\to\mathbb{R}$ defined by $g(x)=x^2+1$ is given by
$$g^4(x)=(((x^2+1)^2+1)^2+1)^2+1.$$
We saw in Exercises 71 and 74 that the cosine function is not a contraction of $\mathbb{R}$ but that its second iterate is a contraction. In fact we shall now see that provided *some iterate* of f is a contraction we still get a fixed-point result similar to the contraction mapping principle.

Theorem 4.3 Let (X,d) be a complete metric space and let $f\colon X\to X$ have the property that, for some N, the iterate f^N is a contraction of X. Then f has a unique fixed point. Furthermore the usual iterative process starting at any $x_1\in X$ and calculating
$$x_1,\quad x_2=f(x_1),\quad x_3=f(x_2),\quad\ldots$$
yields a sequence converging to that unique fixed point.

Proof It is a pleasant surprise that we need not return to Cauchy sequences and the details of the proof of the contraction mapping principle, for our generalisation follows more easily than that. We know that, as f^N is a contraction of the complete space (X,d), f^N has a unique fixed point x_0. Hence

Some extensions 65

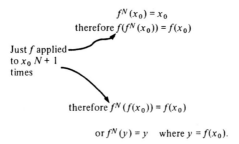

or $f^N(y) = y$ where $y = f(x_0)$.

So y is another fixed point of f^N. But f^N only has one, namely x_0. Hence $f(x_0) = y = x_0$ and x_0 is a fixed point of f. It must be unique, since any point which f fixes clearly remains fixed by f^N.

Finally, to see that
$$x_1, \quad x_2 = f(x_1), \quad x_3 = f(x_2), \quad \ldots$$
converges to x_0, we relabel f^N as g and note that
$$x_{N+1} = f(x_N) = f(f(x_{N-1})) = \cdots = f^N(x_1) = g(x_1).$$
Similarly $x_{N+2} = g(x_2)$, etc. So we now rewrite the sequence $x_1, x_2, x_3, \ldots$ as
$$x_1, x_2, x_3, \ldots, x_N, g(x_1), g(x_2), g(x_3), \ldots, g(x_N),$$
$$g(g(x_1)), g(g(x_2)), \ldots.$$
This is actually a combination of the N sequences
$$x_1, g(x_1), g(g(x_1)), \ldots$$
$$x_2, g(x_2), g(g(x_2)), \ldots$$
$$\vdots$$
$$x_N, g(x_N), g(g(x_N)), \ldots.$$
Each of these is obtained by starting at some point of X and iterating with the contraction g. We know that any such sequence converges to the unique fixed point of g, namely x_0. So each of the N sequences converges to x_0 and hence the combined sequence $x_1, x_2, x_3, \ldots$ converges to x_0, as claimed. □

Exercise 75 Let $f: \mathbb{R} \to \mathbb{R}$ be given by $f(x) = e^{-x}$. Use differentiation to show that f is not a contraction of $\mathbb{R}$, but that f^2 is. Use an iterative technique to find (to three decimal places) the unique fixed point of f, and hence the unique root of $\log x + x = 0$. (Observe that in this iteration the sequence of approximations which you get splits into two parts, one increasing towards the limit and the other decreasing. This resembles the proof of Theorem 4.3 where (when dealing with an Nth iterate) the sequence was split into N parts each converging to the fixed point.)

Exercise 76 Let $C(0, \pi/2)$ be the metric space of continuous real functions $x: [0, \pi/2] \to \mathbb{R}$ together with the usual max metric. Let $f: C(0, \pi/2) \to C(0, \pi/2)$ take the member x of $C(0, \pi/2)$ (which is itself a function) to the member $f(x)$ of $C(0, \pi/2)$ given by

$$(f(x))(t) = \int_0^t (x(u) + u) \sin u \, du.$$

So, for example, if x is the zero function ($x(t) = 0$ for all t), then $f(x)$ is the function given by

$$(f(x))(t) = \int_0^t (0 + u) \sin u \, du = \sin t - t \cos t.$$

Let $x, y \in C(0, \pi/2)$ be given by
$x(t) = -t$, $y(t) = 1 - t$ $(0 \leqslant t \leqslant \pi/2)$.
Find $f(x)$ and $f(y)$ and evaluate $d(x, y)$ and $d(f(x), f(y))$. Deduce that f is not a contraction of $C(0, \pi/2)$.

We have seen that the function f in Exercise 76 is not a contraction: but we now set out to show that some iterate of f *is* a contraction. Note that, for any $x, y \in C(0, \pi/2)$ and $t \in [0, \pi/2]$,

$$(f(x))(t) - (f(y))(t) = \int_0^t (x(u) + u) \sin u \, du - \int_0^t (y(u) + u) \sin u \, du$$

$$= \int_0^t (x(u) - y(u)) \sin u \, du$$

$$\leqslant \int_0^t |x(u) - y(u)| \, du.$$

But $d(x, y)$ is the biggest of all the possible values of $|x(u) - y(u)|$ and so

$$(f(x))(t) - (f(y))(t) \leqslant \int_0^t d(x, y) \, du = t d(x, y).$$

In particular $d(f(x), f(y))$ is the largest of all the values of $|(f(x))(t) - (f(y))(t)|$ and so

Some extensions   67

$$d(f(x), f(y)) \leqslant \max{(td(x, y))} = \frac{\pi}{2} d(x, y), \quad 0 \leqslant t \leqslant \frac{\pi}{2}.$$

Exercise 77 Let f be as above. Use the inequalities just established (at first applied to $f(x)$ and $f(y)$ rather than x and y) to show that

$$(f(f(x)))(t) - (f(f(y)))(t) \leqslant \int_0^t |(f(x))(u) - (f(y))(u)|\, du$$

i.e. $f^2(x)$
$$\leqslant \int_0^t u d(x, y)\, du$$
$$= \frac{t^2}{2} d(x, y),$$

and deduce further that

$$(f^3(x))(t) - (f^3(y))(t) \leqslant \frac{t^3}{6} d(x, y).$$

Hence show that for all $x, y \in C(0, \pi/2)$,

$$d(f(x), f(y)) \leqslant \frac{\pi}{2} d(x, y) \quad \text{(seen above)},$$

$$d(f^2(x), f^2(y)) \leqslant \frac{1}{2}\left(\frac{\pi}{2}\right)^2 d(x, y),$$

and

$$d(f^3(x), f^3(y)) \leqslant \frac{1}{6}\left(\frac{\pi}{2}\right)^3 d(x, y).$$

Deduce that the third iterate of f is a contraction of $C(0, \pi/2)$. (If you find some of these details hard simply move on: there is a similar example worked through in full detail at the beginning of the next section.)

We have therefore seen that if $f: C(0, \pi/2) \to C(0, \pi/2)$ is given by

$$(f(x))(t) = \int_0^t (x(u) + u) \sin u\, du,$$

then the third iterate of f is a contraction of the (complete) metric space $C(0, \pi/2)$. Hence, by Theorem 4.3, f has a unique fixed point; i.e. there is precisely one continuous $x: [0, \pi/2] \to \mathbb{R}$ which satisfies

$$x(t) = (f(x))(t) = \int_0^t (x(u) + u) \sin u\, du$$

for all $t \in [0, \pi/2]$. So the 'integral equation'

$$x(t) = \int_0^t (x(u) + u) \sin u \, du \quad \left(0 \leq t \leq \frac{\pi}{2}\right)$$

has a unique solution. But *differential* equations are much more common and useful: what is the relevance of the fixed point of f to a differential equation? We see that the integral equation (left-hand box) is equivalent to a differential equation (right-hand box):

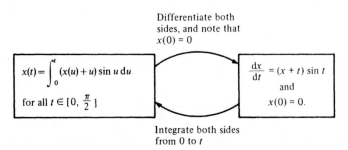

We will meet this again in the next section and it is only included here as a preview. But let us summarise what we have seen in this example. The solutions of the differential equation

$$\frac{dx}{dt} = (x + t) \sin t \quad \text{subject to } x(0) = 0$$

are equivalent to solutions of

$$x(t) = \int_0^t (x(u) + u) \sin u \, du.$$

These, in turn, are the fixed points of a function $f: C(0, \pi/2) \to C(0, \pi/2)$ for which some iterate is a contraction. Hence f has a unique fixed point and the differential equation has a unique solution. We shall return to this topic much more fully in the next section.

That completes our generalisation of the contraction mapping principle to iterates: our second extension involves weakening the requirement that $d(f(x), f(y)) \leq k d(x, y)$ *for some $k < 1$*. We saw by means of examples in the first section of this chapter that the condition $d(f(x), f(y)) < d(x, y)$ for all $x \neq y$ was *not* enough to ensure that f has a fixed point. Our counter-example concerned $f(x) = x + 1/x$ for $x \in [1, \infty[$ where, in a very informal sense, the fixed point was 'at infinity' and therefore not allowed. For our second extension of the

Some extensions

contraction mapping principle we shall see now that if we restrict attention to compact spaces then the slightly weaker condition that f reduces distances (rather than reduces them by at least a factor of $k<1$) is sufficient to ensure a unique fixed point. Furthermore our iterative process will again lead to that fixed point. Those readers who chose to pass quickly over compactness in the previous chapter can easily miss out this theorem.

Theorem 4.4 Let (X, d) be a compact metric space (i.e. X is itself compact) and let $f: X \to X$ satisfy

$$d(f(x), f(y)) < d(x, y)$$

for all $x \neq y$ in X. Then f has a unique fixed point x_0. Furthermore if $x_1 \in X$ and $x_2 = f(x_1)$, $x_3 = f(x_2)$, and so on, then the sequence $x_1, x_2, x_3, \ldots$ converges to x_0.

Proof For each $x \in X$ the distance $d(x, f(x))$ is a real number. Define $F: X \to \mathbb{R}$ by $F(x) = d(x, f(x))$. By the triangle inequality, for any x and y in X

$$d(x, f(x)) \leq d(x, y) + d(y, f(y)) + d(f(y), f(x)),$$

and so

$$|F(x) - F(y)| = |d(x, f(x)) - d(y, f(y))|$$
$$\leq d(x, y) + d(f(y), f(x))$$
$$\leq 2d(x, y)$$

(by the given distance-reducing property of f).

So, as in Exercise 54, the set

$$F(X) = \{F(x): x \in X\} = \{d(x, f(x)): x \in X\}$$

has a least element, α, say, equal to $d(x_0, f(x_0))$. The exercise will now show us that $\alpha = 0$ and hence that $x_0 = f(x_0)$.

Exercise 78 Show that if $\alpha \neq 0$ then the distance from $f(x_0)$ to $f(f(x_0))$ is less than α. Deduce that f has a fixed point.

Hence f has a fixed point x_0 and (by Exercise 66) this point is unique. The remaining question, crucial to our approach, is whether choosing $x_1 \in X$ and evaluating $x_2 = f(x_1)$, $x_3 = f(x_2)$, and so on, will give a sequence converging to x_0. So let $x_1, x_2, x_3, \ldots$ be constructed in that way. Then

$$d(x_n, x_0) = d(f(x_{n-1}), f(x_0)) \leq d(x_{n-1}, x_0)$$

and so the sequence of real numbers
$$d(x_1, x_0), d(x_2, x_0), d(x_3, x_0), \ldots$$
is decreasing and hence convergent to some number $\beta \geq 0$. We aim to show that $\beta = 0$ (so that
$$d(x_1, x_0), d(x_2, x_0), d(x_3, x_0), \ldots \to 0;$$
i.e. $x_1, x_2, x_3, \ldots \to x_0$ as claimed).

By the compactness of (X, d) the sequence $x_1, x_2, x_3, \ldots$ certainly has a convergent subsequence
$$x_{k_1}, x_{k_2}, x_{k_3}, \ldots \to y \quad \text{(say)}.$$
Therefore
$$d(x_{k_1}, x_0), d(x_{k_2}, x_0), d(x_{k_3}, x_0), \ldots \to d(y, x_0)$$
since $d(x_n, x_0)$ lies between $d(y, x_0) - d(y, x_n)$ and $d(y, x_0) + d(y, x_n)$ (the triangle inequality). But
$$d(x_1, x_0), d(x_2, x_0), d(x_3, x_0), \ldots \to \beta$$
and so $d(y, x_0) = \beta$. We have shown therefore that *any* convergent subsequence of $x_1, x_2, x_3, \ldots$ has its limit at a distance β from x_0. But as
$$x_{k_1}, x_{k_2}, x_{k_3}, \ldots \to y$$
and f reduces distances it follows that
$$f(x_{k_1}), f(x_{k_2}), f(x_{k_3}), \ldots \to f(y);$$
$$\| \qquad \| \qquad \|$$
$$x_{k_1+1}, x_{k_2+1}, x_{k_3+1}, \ldots \to f(y).$$
So we have another convergent subsequence of $x_1, x_2, x_3, \ldots$ and by the above comments *its* limit must be distance β from x_0; i.e. $d(f(y), x_0) = \beta$. Hence
$$d(f(y), f(x_0)) = d(f(y), x_0) = \beta = d(y, x_0)$$
and so by the fact that f actually reduces distances between distinct points it follows that $x_0 = y$ and $\beta = 0$.

Therefore β is 0, the sequence of $d(x_n, x_0)$s does converge to 0, and $x_1, x_2, x_3, \ldots$ does converge to x_0, the fixed point of f, as required. □

Exercise 79 Let $f: [-\pi, \pi] \to [-\pi, \pi]$ be given by $f(x) = \sin x + 1$. Show that $|f'(x)| < 1$ for all $x \in \,]-\pi, \pi[$. Use the technique of the proof of Theorem 4.2 to deduce that
$$|f(x) - f(y)| < |x - y|$$

Differential equations

for all $x \neq y$ in $[-\pi, \pi]$. What can you deduce about the equation
$$\sin x - x + 1 = 0?$$
Find a root of the equation (to three decimal places, say).

During our development of the contraction mapping principle we have seen applications to real equations (Exercises 61, 62, 71, 72, 73, 75 and others), simultaneous equations (Exercises 63, 64 and 67), linear algebra (Exercises 69 and 70) and differential equations (Exercises 76 and 77). We now examine that last application in much more detail.

4.5 Differential equations

The contraction mapping principle will enable us to deduce a general result about the solutions of a differential equation of the form $dx/dt = F(x,t)$. But in order to understand the result (and proof) we begin by working through one example in its full detail.

Let us start therefore by looking at the differential equation
$$\frac{dx}{dt} = (x + t^2) e^{t-1}$$
where we want a solution x defined for $t \in [1, 3]$ and which satisfies the initial condition $x(1) = 4$. Integrating the equation from 1 to t and using the fact that $x(1) = 4$ gives
$$x(t) = 4 + \int_1^t (x(u) + u^2) e^{u-1} \, du \quad (1 \leq t \leq 3).$$
(Indeed, as we remarked when considering the differential equation in the previous section, this integral equation ensures that $x(1) = 4$ and on differentiation we get back to
$$\frac{dx}{dt} = (x + t^2) e^{t-1}.$$
So the integral equation is *equivalent to* the differential equation together with the initial condition.)

Therefore we can switch attention to the integral equation. For each $x \in C(1, 3)$ let $f(x)$ be the function in $C(1, 3)$ given by
$$(f(x))(t) = 4 + \int_1^t (x(u) + u^2) e^{u-1} \, du \quad (1 \leq t \leq 3).$$

Then the solutions of the integral equation are precisely those functions $x \in C(1, 3)$ for which $f(x) = x$; i.e. the fixed points of f. We shall investigate whether $f: C(1, 3) \to C(1, 3)$ (or some iterate of it) is a contraction.

For each $x, y \in C(1, 3)$ and $t \in [1, 3]$

$$(f(x))(t) - (f(y))(t) = \left[4 + \int_1^t (x(u) + u^2) e^{u-1} \, du\right]$$
$$- \left[4 + \int_1^t (y(u) + u^2) e^{u-1} \, du\right]$$
$$= \int_1^t (x(u) - y(u)) e^{u-1} \, du.$$

Recall that $d(x, y)$ is the largest of all the values of $|x(u) - y(u)|$ for $1 \leqslant u \leqslant 3$, and also the biggest value of e^{u-1} for $u \in [1, 3]$ is e^2. Therefore

$$(f(x))(t) - (f(y))(t)$$
$$= \int_1^t (x(u) - y(u)) e^{u-1} \, du \leqslant \int_1^t |x(u) - y(u)| \, e^2 \, du$$
$$\leqslant \int_1^t d(x, y) \, e^2 \, du = d(x, y) \, e^2 (t - 1).$$

If we apply the first part of this process to $f(x)$ and $f(y)$ rather than x and y we get

$$(f^2(x))(t) - (f^2(y))(t) \leqslant \int_1^t |(f(x))(u) - (f(y))(u)| \, e^2 \, du.$$

But we have also established that

$$(f(x))(t) - (f(y))(t) \leqslant d(x, y) \, e^2 (t - 1)$$

and so

$$|(f(x))(u) - (f(y))(u)| = \begin{cases} (f(x))(u) - (f(y))(u) \\ \text{or} \\ (f(y))(u) - (f(x))(u) \end{cases}$$
$$\leqslant d(x, y) \, e^2 (u - 1).$$

Hence

$$(f^2(x))(t) - (f^2(y))(t) \leqslant \int_1^t |(f(x))(u) - (f(y))(u)| \, e^2 \, du$$

Differential equations

$$\leq \int_1^t d(x, y) \, e^4 (u-1) \, du$$

$$= d(x, y) e^4 \frac{(t-1)^2}{2}.$$

Exercise 80 Continue the above process to deduce that

$$(f^3(x))(t) - (f^3(y))(t) \leq d(x, y) \, e^6 \frac{(t-1)^3}{6}$$

and

$$(f^4(x))(t) - (f^4(y))(t) \leq d(x, y) \, e^8 \frac{(t-1)^4}{24}.$$

State the general result concerning $(f^N(x))(t) - (f^N(y))(t)$.

Since, for $1 \leq t \leq 3$,

$$(f(x))(t) - (f(y))(t) \leq d(x, y) \, e^2 (t-1) \leq 2 \, e^2 d(x, y)$$

it follows that

$$d(f(x), f(y)) = \max_{1 \leq t \leq 3} |(f(x))(t) - (f(y))(t)| \leq 2 \, e^2 d(x, y).$$

However, $2e^2 \approx 15$ and so the fact that $d(f(x), f(y)) \leq 2e^2 d(x, y)$ is not much help in showing whether or not f is a contraction! Similarly we saw above that for $1 \leq t \leq 3$

$$(f^2(x))(t) - (f^2(y))(t) \leq d(x, y) \, e^4 \frac{(t-1)^2}{2} \leq \frac{2^2}{2} \, e^4 d(x, y) = 2e^4 d(x, y)$$

and so $d(f^2(x), f^2(y)) \leq 2e^4 d(x, y)$: but again $2e^4 \approx 109$ is much too large to be of any help.

Exercise 81 Use the results of the previous exercise to show that

$$d(f^3(x), f^3(y)) \leq \frac{2^3 e^6}{6} d(x, y)$$

$$d(f^4(x), f^4(y)) \leq \frac{2^4 e^8}{24} d(x, y)$$

and, in general,

$$d(f^N(x), f^N(y)) \leq \frac{2^N e^{2N}}{N!} d(x, y).$$

How does Exercise 81 help us to establish that some iterate of f is a contraction? We look in detail at the factors by which the iterates change distances:

	'factor' k	
$d(f(x), f(y)) \leqslant 2e^2 d(x, y)$	$2e^2 \approx 15$	in going from the first to the second the factor multiplies by $2e^2/2$
$d(f^2(x), f^2(y)) \leqslant \dfrac{2^2}{2} e^4 d(x, y)$	$\dfrac{2^2}{2} e^4 \approx 109$	in going from the second to the third the factor multiplies by $2e^2/3$
$d(f^3(x), f^3(y)) \leqslant \dfrac{2^3}{3!} e^6 d(x, y)$	$\dfrac{2^3}{3!} e^6 \approx 538$	in going from the third to the fourth the factor multiplies by $2e^2/4$
$d(f^4(x), f^4(y)) \leqslant \dfrac{2^4}{4!} e^8 d(x, y)$	$\dfrac{2^4}{4!} e^8 \approx 1987$	
$\vdots$	$\vdots$	$\vdots$
$d(f^{13}(x), f^{13}(y)) \leqslant \dfrac{2^{13}}{13!} e^{26} d(x, y)$	$\dfrac{2^{13}}{13!} e^{26} \approx 257493$	multiplied by $2e^2/14$ (>1) so the factors are still rising
$d(f^{14}(x), f^{14}(y)) \leqslant \dfrac{2^{14}}{14!} e^{28} d(x, y)$	$\dfrac{2^{14}}{14!} e^{28} \approx 271805$	multiplied by $2e^2/15$ (<1) so the factors, at last, start to decrease
$d(f^{15}(x), f^{15}(y)) \leqslant \dfrac{2^{15}}{15!} e^{30} d(x, y)$	$\dfrac{2^{15}}{15!} e^{30} \approx 267784$	
$\vdots$	$\vdots$	
$d(f^{36}(x), f^{36}(y)) \leqslant \dfrac{2^{36}}{36!} e^{72} d(x, y)$	$\dfrac{2^{36}}{36!} e^{72} \approx 3.43$	
$d(f^{37}(x), f^{37}(y)) \leqslant \dfrac{2^{37}}{37!} e^{74} d(x, y)$	$\dfrac{2^{37}}{37!} e^{74} \approx 1.31$	
$d(f^{38}(x), f^{38}(y)) \leqslant \dfrac{2^{38}}{38!} e^{76} d(x, y)$	$\dfrac{2^{38}}{38!} e^{76} \approx 0.53$	

Incidentally $e^2 = 1 + 2 + \dfrac{2^2}{2!} + \dfrac{2^3}{3!} + \cdots$ which is another way of seeing that the terms $2^n/n!$ tend to zero.

Hence $d(f^{38}(x), f^{38}(y)) \leqslant k d(x, y)$ where k is approximately 0.53; i.e. the 38th iterate of f is a contraction! Therefore f has a unique fixed point and (if you haven't forgotten by now) this means that the original integral equation has a unique solution. We have therefore (rather tortuously) seen that the differential equation

Differential equations

$$\frac{dx}{dt} = (x+t^2)e^{t-1} \quad (1 \leq t \leq 3), \quad x(1)=4$$

has a unique solution.

Of course we shall not have to go through this process for each differential equation! Our next theorem will cover all equations of this type in one go: the advantage of having worked through this example at length is that, with luck, the proof will now seem natural and easy. We make one more comment before proceeding to that general result. The contraction mapping principle, apart from establishing the existence of unique fixed points, enables us to find the fixed point by repeatedly applying the function. In this particular application reapplying the function will mean integrating a succession of functions. This is generally a daunting task even if the integration is elementary. For example, to solve the differential equation

$$\frac{dx}{dt} = (x+t^2)e^{t-1}, \quad x(1)=4$$

considered earlier would require iterating with f given by

$$(f(x))(t) = 4 + \int_1^t (x(u)+u^2)e^{u-1}\,du.$$

Starting with an initial guess of $x_1(t)=0$, say, each stage would involve integrals of terms of the form $u^n e^{mu+p}$: these are reasonably elementary to integrate but remember that the 38th iterate of f was the contraction, so we would have to perform 38 iterations before we could be sure that our function was any closer to the root than x_1! An alternative is to use a computer to calculate successive terms by numerical integration. In general, though, our method is a delightful way of showing the existence and uniqueness of solutions of differential equations without actually finding them. But before proceeding to the general result we include one exercise where the solution of the differential equation can actually be found by iteration. That is not to say, of course, that it is the most practical way of solving the equation.

Exercise 82 The solutions of the differential equation

$$\frac{dx}{dt} = (x+t)t \quad (0 \leq t \leq 1), \quad x(0)=0$$

are the fixed points of the function $f\colon C(0,1) \to C(0,1)$ given by

$$(f(x))(t) = \int_0^t (x(u)+u)u\,du.$$

Show that f is a contraction. Hence the differential equation has a unique solution. To find the solution let x_1 be given by $x_1(t) = 0$ for all $t \in [0,1]$. Let $x_2 = f(x_1)$, $x_3 = f(x_2)$ and so on. Show that $x_2(t) = t^3/3$ and $x_3(t) = (t^3/3) + (t^5/15)$. Convince yourselves by evaluating a few more terms in the sequence that $x_1, x_2, x_3, \ldots$ converges to the functions given by

$$x(t) = \frac{t^3}{3} + \frac{t^5}{5\cdot 3} + \frac{t^7}{7\cdot 5\cdot 3} + \frac{t^9}{9\cdot 7\cdot 5\cdot 3} + \cdots.$$

This, in series form, is therefore the required unique solution of the differential equation.

Theorem 4.5 Let $F\colon \mathbb{R} \times [a,b] \to \mathbb{R}$; i.e. F is a real function of two variables such that $F(x,t)$ is defined for all $x \in \mathbb{R}$ and $t \in [a,b]$. Assume that F is continuous and that there exists a fixed real number L with

$$|F(x,t) - F(y,t)| \leq L|x-y| \quad \begin{cases} \text{This is called a} \\ \text{`Lipschitzian condition'} \end{cases}$$

for all $x, y \in \mathbb{R}$ and $t \in [a,b]$. Then the differential equation

$$\frac{dx}{dt} = F(x,t),$$

subject to an initial condition of the type $x(\alpha) = \beta$, has a unique solution.

Proof (We will assume that the initial condition is $x(a) = \beta$: there is no real loss in generality, but it makes the proof a little easier.) Having seen two particular examples worked through ($F(x,t) = (x+t)\sin t$ in Exercise 76, etc., and $F(x,t) = (x+t^2)e^{t-1}$ at the beginning of this section), the reader will probably know exactly how the proof will proceed. Define $f\colon C(a,b) \to C(a,b)$ by

$$(f(x))(t) = \beta + \int_a^t F(x(u), u)\,du.$$

Then the required solutions of the differential equations are precisely the fixed points of f.

Now, for each $x, y \in C(a,b)$ and $t \in [a,b]$,

Differential equations

$$(f(x))(t) - (f(y))(t)$$
$$= \left[\beta + \int_a^t F(x(u), u)\, du\right] - \left[\beta + \int_a^t F(y(u), u)\, du\right]$$
$$= \int_a^t [F(x(u), u) - F(y(u), u)]\, du$$
$$\leq \int_a^t L|x(u) - y(u)|\, du \quad \text{(by the Lipschitzian condition)}$$
$$\leq \int_a^t L d(x, y)\, du = L d(x, y)(t - a).$$

Hence, in a similar fashion, but with $f(x)$ and $f(y)$ rather than x and y,

$$(f^2(x))(t) - (f^2(y))(t) \leq \int_a^t L|(f(x))(u) - (f(y))(u)|\, du.$$

But we have already shown that

$$(f(x))(t) - (f(y))(t) \leq L d(x, y)(t - a)$$

and so

$$(f^2(x))(t) - (f^2(y))(t) \leq \int_a^t L|(f(x))(u) - (f(y))(u)|\, du$$
$$\leq \int_a^t L^2 d(x, y)(u - a)\, du$$
$$= L^2 d(x, y) \frac{(t - a)^2}{2}.$$

Continuing in this way it is easy to deduce that in general

$$(f^N(x))(t) - (f^N(y))(t) \leq L^N d(x, y) \frac{(t - a)^N}{N!}$$
$$\leq L^N d(x, y) \frac{(b - a)^N}{N!} \quad \text{(since } t \in [a, b]\text{)}.$$

Hence for each positive integer N

$$d(f^N(x), f^N(y)) \leq \frac{L^N (b - a)^N}{N!} d(x, y).$$

Now as $N \to \infty$ the numbers $L^N (b - a)^N / N!$ tend to zero. For, as we saw in the earlier example, as soon as N is larger than the fixed number $L(b - a)$ the terms $L^N (b - a)^N / N!$ start to fall and eventually to decrease dramatically. Hence there exists some $L^{N_0}(b - a)^{N_0}/N_0!$

78 The contraction mapping principle

less than 1, and so the N_0th iterate of f is then a contraction. Thus, by Theorem 4.3, f has a unique fixed point; and the differential equation has a unique solution. □

In practice, finding whether there exists such an L as in Theorem 4.5 is straightforward, as the next result and exercises show.

Theorem 4.6 Let $F: \mathbb{R} \times [a,b] \to \mathbb{R}$; i.e. F is a real function of two variables such that $F(x,t)$ is defined for all $x \in \mathbb{R}$ and $t \in [a,b]$. Assume that F is continuous, that F can be partially differentiated with respect to x (i.e. with t fixed) and that $\partial F/\partial x$ is bounded throughout $\mathbb{R} \times [a,b]$. Then the differential equation

$$\frac{dx}{dt} = F(x,t),$$

subject to an initial condition of the type $x(\alpha) = \beta$, has a unique solution.

Proof Assume that $|\partial F/\partial x| \leq L$ for all $x \in \mathbb{R}$ and $t \in [a,b]$. Then let us fix t and define a function G of x alone by

$$G(x) = F(x, \underset{\substack{\uparrow \\ \text{fixed}}}{t}).$$

It follows that G is differentiable and, by the mean value theorem we recalled earlier, for any distinct x and y in $\mathbb{R}$

$$G(x) - G(y) = G'(c)(x-y)$$

for some c between x and y. Hence

$$|F(x,t) - F(y,t)| = |G(x) - G(y)| = |G'(c)||x-y|$$
$$= \underset{\substack{\uparrow \\ \text{evaluated at} \\ (c,t)}}{\left|\frac{\partial F}{\partial x}\right|} |x-y| \leq L|x-y|.$$

Thus F satisfies a Lipschitzian condition as in Theorem 4.5 and the uniqueness of the solution of the differential equation follows from that result. □

Exercise 83 Show that for any positive real number T and any real number x

Differential equations

$$\left|\frac{2x}{(T^2+x^2)^2}\right| \leqslant \left|\frac{2x}{T^2+x^2}\right| \frac{1}{T^2} \leqslant \frac{1}{T^3}.$$

Now let

$$F(x,t) = \frac{t}{2^t + x^2}$$

for $x \in \mathbb{R}$ and $t \in [-10, 10]$. Use the above inequalities with $T = 2^{t/2}$ to show that

$$\left|\frac{\partial F}{\partial x}\right| \leqslant 10 \cdot 2^{15}$$

throughout $\mathbb{R} \times [-10, 10]$. It follows, therefore, from Theorem 4.6, that the differential equation

$$\frac{dx}{dt} = \frac{t}{2^t + x^2}, \quad x(0) = 1$$

has a unique solution defined on $[-10, 10]$. Show that the same conclusion can be drawn for any interval $[-n, n]$. Do you see why this means that the differential equation

$$\frac{dx}{dt} = \frac{t}{2^t + x^2}, \quad x(0) = 1$$

has a unique solution x defined for $t \in \mathbb{R}$?

Exercise 84 Let $[a, b]$ be an interval contained in $]0, \infty[$. Show that for $t \in [a, b]$ and $x \in \mathbb{R}$

$$\frac{e^x}{(t+e^x)^2} \leqslant \frac{1}{t} \leqslant \frac{1}{a}.$$

Use Theorem 4.6 to show that the differential equation

$$\frac{dx}{dt} = \frac{1}{t + e^x}$$

subject to an initial condition $x(\alpha) = \beta$ has a unique solution x defined for $t \in [a, b]$. Can you see why this means that there is a unique such solution x defined for $t \in]0, \infty[$?

There is an extension of the above approach to simultaneous differential equations and hence to higher-order linear differential equations: the reader who has had enough of this topic can turn straight to the next section.

We saw in Exercise 64 that in order to solve the simultaneous

equations
$$x_1 = \tfrac{1}{2}\cos x_2 \quad \text{and} \quad x_2 = \tfrac{1}{2}\sin x_1 + 1$$
we could consider the function $f: \mathbb{R}^2 \to \mathbb{R}^2$ given by $f(x_1, x_2) = (\tfrac{1}{2}\cos x_2, \tfrac{1}{2}\sin x_1 + 1)$ and look for its fixed points. In other words, to solve two real equations simultaneously we would consider pairs of members of $\mathbb{R}$. Similarly, in Exercise 67, in order to solve three real equations simultaneously we used the metric space of triples of members of $\mathbb{R}$. So if we want to consider a set of simultaneous differential equations (with solutions on $[a, b]$, say) perhaps we ought to consider a metric space consisting of lists of members of $C(a, b)$. Needless to say the approach can get a little technical and so we give only a broad outline (leaving most of the work to you!).

Let us see first why the solutions of simultaneous differential equations are relevant to the solutions of linear differential equations of higher order.

Exercise 85 Show that given any solution (for $0 \leq t \leq 1$, say) of the simultaneous differential equations
$$\frac{dx_1}{dt} = x_2, \quad x_1(0) = 1 \quad \text{and} \quad \frac{dx_2}{dt} = e^t - x_1 - 2tx_2, \quad x_2(0) = 2$$
the function x_1 is a solution of
$$\frac{d^2x_1}{dt^2} + 2t\frac{dx_1}{dt} + x_1 = e^t, \quad x_1(0) = 1, \quad \frac{dx_1}{dt}(0) = 2.$$
Conversely, show that if x_1 is a solution of this second-order equation, then the pair x_1 and $x_2 = dx_1/dt$ satisfy the simultaneous equations.

If we rewrite the equations
$$\frac{dx_1}{dt} = x_2 \; (= F_1(x_1, x_2, t), \text{ say}), \quad x_1(0) = 1$$
and
$$\frac{dx_2}{dt} = e^t - x_1 - 2tx_2 \; (= F_2(x_1, x_2, t), \text{ say}), \quad x_2(0) = 2$$
in condensed form as
$$\frac{d\mathbf{x}}{dt} = \mathbf{F}(\mathbf{x}, t), \quad \mathbf{x}(0) = (1, 2),$$
then there is a close superficial resemblance to the differential

Differential equations

equations considered in Theorem 4.5. Imitating the proof of that result, perhaps we should set up a metric space (X, d) of *pairs* of continuous functions and then define a relevant $f: X \to X$ where fixed points are the required solutions. We could then try to show that some iterate of f is a contraction.

Exercise 86 Let X be the set of pairs $\mathbf{x} = (x_1, x_2)$ of continuous real functions defined on $[0, 1]$. Define d by

$$d(\mathbf{x}, \mathbf{y}) = d((x_1, x_2), (y_1, y_2))$$

$$= \max_{0 \leq t \leq 1} |x_1(t) - y_1(t)| + \max_{0 \leq t \leq 1} |x_2(t) - y_2(t)|$$

for $\mathbf{x}, \mathbf{y} \in X$. Show that (X, d) is a metric space and is complete.

Exercise 87 Let X be as in the previous exercise and define $f: X \to X$ by $f(\mathbf{x}) = f(x_1, x_2) = (y_1, y_2)$ where

$$y_1(t) = 1 + \int_0^t x_2(u) \, du$$

and

$$y_2(t) = 2 + \int_0^t (e^u - x_1(u) - 2ux_2(u)) \, du.$$

Show that, given any fixed point (x_1, x_2) of f, the function x_1 is a solution of the differential equation

$$\frac{d^2 x_1}{dt^2} + 2t \frac{dx_1}{dt} + x_1 = e^t, \quad x_1(0) = 1, \quad \frac{dx_1}{dt}(0) = 2.$$

Conversely, show that if x_1 is a solution of this differential equation then $\mathbf{x} = (x_1, dx_1/dt)$ is a fixed point of f.

In general, to consider the solutions of an nth-order linear differential equation, we shall rewrite it as n simultaneous first-order differential equations. We shall then show that these solutions coincide with the fixed points of a function on the metric space of n-tuples of continuous functions. The result is stated below, but only a very broad outline of the proof is given. It generalises the idea, seen in introductory courses, that a second-order linear differential equation with two initial conditions has a unique solution.

Theorem 4.7 Let $h, g_1, g_2, \ldots, g_n: [a, b] \to \mathbb{R}$ be continuous functions. Then the differential equation

82 The contraction mapping principle

$$\frac{d^n x_1}{dt^n} + g_1(t)\frac{d^{n-1}x_1}{dt^{n-1}} + g_2(t)\frac{d^{n-2}x_1}{dt^{n-2}} + \cdots + g_n(t)x_1(t) = h(t)$$

subject to initial conditions

$$x_1(\alpha) = \beta_1, \quad \frac{dx_1}{dt}(\alpha) = \beta_2, \ldots, \quad \frac{dx_1^{n-1}}{dt^{n-1}}(\alpha) = \beta_n$$

has a unique solution.

Outline proof The solutions of the required differential equation are equivalent to the solutions of the simultaneous equations

$$\frac{dx_1}{dt} = x_2 \ (= F_1(x_1, \ldots, x_n, t), \text{ say}), \quad x_1(\alpha) = \beta_1,$$

$$\frac{dx_2}{dt} = x_3 \ (= F_2(x_1, \ldots, x_n, t), \text{ say}), \quad x_2(\alpha) = \beta_2,$$

$$\vdots$$

$$\frac{dx_n}{dt} = h(t) - g_1(t)x_n - g_2(t)x_{n-1} - \cdots - g_n(t)x_1$$

$$(= F_n(x_1, \ldots, x_n, t), \text{ say}), \quad x_n(\alpha) = \beta_n.$$

We abbreviate these n equations to

$$\frac{d\mathbf{x}}{dt} = \mathbf{F}(\mathbf{x}, t), \quad \mathbf{x}(\alpha) = \boldsymbol{\beta}$$

where $\mathbf{x} = (x_1, \ldots, x_n)$, $\mathbf{F}$ stands for the n functions $(F_1, \ldots, F_n)$, and $\boldsymbol{\beta} = (\beta_1, \ldots, \beta_n)$.

So now let X be the set of n-tuples of continuous real functions defined on $[a, b]$. Define d by

$$d(\mathbf{x}, \mathbf{y}) = d((x_1, \ldots, x_n), (y_1, \ldots, y_n))$$

$$= \max_{a \leq t \leq b} |x_1(t) - y_1(t)| + \cdots + \max_{a \leq t \leq b} |x_n(t) - y_n(t)|.$$

Then (X, d) is a complete metric space.

Now define $f: X \to X$ by $f(\mathbf{x}) = f(x_1, \ldots, x_n) = (y_1, \ldots, y_n) = \mathbf{y}$, where

$$y_i(t) = \beta_i + \int_\alpha^t F_i(x_1(u), \ldots, x_n(u), u)\, du.$$

Then some iterate of f is a contraction. When trying to prove the corresponding result for a single equation in the proof of Theorem 4.5 we needed the fact that

$F(x, t) - F(y, t) \leq L|x - y|.$

The fact that some iterate of our new f is a contraction will follow similarly from the fact that we can find a fixed number L with distance from $F(x, t)$ to $F(y, t) \leq L \cdot$ (distance from x to y)

(in $\mathbb{R}^n$) \hspace{4em} (in $\mathbb{R}^n$)

For the distance from $F(x,t)$ to $F(y,t)$ is

$$\{[F_1(x,t) - F_1(y,t)]^2 + [F_2(x,t) - F_2(y,t)]^2 + \cdots + [F_n(x,t) - F_n(y,t)]^2\}^{1/2}$$

$$= \{(x_2 - y_2)^2 + (x_3 - y_3)^2 + \cdots + [g_1(t)(x_n - y_n) + \cdots + g_n(t)(x_1 - y_1)]^2\}^{1/2}$$

$$\leq L\{(x_1 - y_1)^2 + (x_2 - y_2)^2 + \cdots + (x_n - y_n)^2\}^{1/2}$$

where $L = (2n - 1)^{1/2} \max_{a \leq t \leq b} \{g_1(t), g_2(t), \ldots, g_n(t), 1\}$. We leave the interested reader to check this fact.

Hence the function F does satisfy a Lipschitzian condition and it can be shown, in a fashion similar to that in the proof of Theorem 4.5, that some iterate of f is a contraction. Hence f has a unique fixed point and the original differential equation has a unique solution. □

4.6 The implicit function theorem

Our final application of the contraction mapping principle is to a classical theorem from analysis. We introduce this theorem by means of an example.

Exercise 88 Show that the equation

$$\frac{\pi}{4} + x + \tan^{-1} x = 0$$

has a unique solution. (This can be done in a naive way considering the signs of the left-hand side and also showing that the expression is increasing, or by rearranging the equation into the form $x = f(x)$ and using our earlier theory.) Show similarly that for any *fixed* $t > -1$ there is a unique real number x with

$$\tan^{-1} t + x + \tan^{-1}(xt) = 0.$$

The result of Exercise 88 is that the equation

$$\tan^{-1} t + x + \tan^{-1}(xt) = 0$$

84 The contraction mapping principle

uniquely defines a function x in terms of t for all $t > -1$. We say that x is defined *implicitly* in terms of t (rather than *explicitly* when x is given in the form $x = f(t)$). But in Exercise 88 we only showed that x was defined for each individual t and we learnt nothing of the overall behaviour of x: in fact it turns out to be a continuous function. We shall prove this using the following result.

Theorem 4.8 (The implicit function theorem.) Let $F: \mathbb{R} \times [a, b] \to \mathbb{R}$; i.e. F is a real function of two variables such that $F(x, t)$ is defined for all $x \in \mathbb{R}$ and $t \in [a, b]$. Assume that F is continuous and that there exist numbers m and M with

$$0 < m \leqslant \frac{\partial F}{\partial x} \leqslant M$$

for all $x \in \mathbb{R}$ and $t \in [a, b]$. Then there exists a unique continuous function $x: [a, b] \to \mathbb{R}$ such that $F(x(t), t) = 0$ for all $t \in [a, b]$; i.e. the equation $F(x, t) = 0$ does implicitly define a unique continuous function x in terms of t.

Proof Let $f: C(a, b) \to C(a, b)$ take the function $x \in C(a, b)$ to the function $f(x)$ given by

$$(f(x))(t) = x(t) - \frac{F(x(t), t)}{M} \quad (a \leqslant t \leqslant b).$$

We claim that f is a contraction of $C(a, b)$. For if $x, y \in C(a, b)$ and $t \in [a, b]$ then, as we saw in the proof of Theorem 4.6, the mean value theorem shows that

$$F(x, t) - F(y, t) = \frac{\partial F}{\partial x} (x - y).$$

↑
evaluated at (c, t)
for some c between
x and y

Hence

$$(f(x))(t) - (f(y))(t) = \left[x(t) - \frac{F(x(t), t)}{M} \right] - \left[y(t) - \frac{F(y(t), t)}{M} \right]$$

$$= (x(t) - y(t)) - \frac{1}{M} (F(x(t), t) - F(y(t), t))$$

The implicit function theorem

$$= (x(t) - y(t)) - \frac{1}{M} \cdot \frac{\partial F}{\partial x} \cdot (x(t) - y(t))$$

↑
evaluated at (c, t)
for some c between
$x(t)$ and $y(t)$

$$= \left(1 - \frac{1}{M} \cdot \frac{\partial F}{\partial x}\right)(x(t) - y(t)) \leqslant \left(1 - \frac{m}{M}\right)(x(t) - y(t)).$$

Therefore, with the usual metric on $C(a, b)$,

$$d(f(x), f(y)) = \max_{a \leqslant t \leqslant b} |f(x)(t) - (f(y))(t)|$$

$$\leqslant \max_{a \leqslant t \leqslant b} \left\{\left|1 - \frac{m}{M}\right| |x(t) - y(t)|\right\}$$

$$\leqslant \left(1 - \frac{m}{M}\right) d(x, y).$$

Since $1 - m/M < 1$ it follows that f is a contraction of $C(a, b)$ and therefore has a unique fixed point, x, say. Thus for all $t \in [a, b]$

$$x(t) = (f(x))(t) = x(t) - \frac{F(x(t), t)}{M}$$

and

$$F(x(t), t) = 0.$$

It follows that x is a continuous solution of $F(x, t) = 0$. Its uniqueness follows from the fact that any such solution must be a fixed point of f. □

Exercise 89 Show that the equation

$$4x - 3 \cos x = 1 - 2t \cos t$$

uniquely defines a continuous function x in terms of t, for t in any interval $[a, b]$. Why does this mean that a continuous x is uniquely defined for $t \in \mathbb{R}$?

Exercise 90 Show that the equation

$$\tan^{-1} t + x + \tan^{-1}(xt) = 0$$

uniquely defines a continuous function for $t \in [a, b]$ where $-1 < a < 0 < b$. (It might help to note that for such an a, b and t

$$a \leqslant \frac{a}{1+x^2t^2} \leqslant \frac{t}{1+x^2t^2} \leqslant \frac{b}{1+x^2t^2} \leqslant b.)$$

Why does this mean that a continuous x is uniquely defined for $t \in \,]-1, \infty[\,$?

4.7 Conclusion

Those readers who are not interested in pure mathematical abstraction will probably not wish to read the final chapter which generalises real analysis and at the same time shows us the underlying reasons why it works. It is for such readers that the current approach has evolved, with its main theme of iteration by contraction. Only the concepts necessary for that approach were introduced and they were motivated and explained with that approach in mind. But it would be wrong for the reader to think that this was the only use of metric spaces. This subject, as part of the area of mathematics known as 'functional analysis', has a large variety of uses in both pure and applied mathematics. It just happens that contractions are capable of being understood without a great deal of pure mathematical background.

However, I hope that most readers will feel that, having worked hard to understand the abstract ideas of metrics, closed sets, completeness and compactness, it would be a pity to stop without seeing their relevance to analysis. So we now include a final chapter with the subject looked at in a more traditional way.

5

What makes analysis work?

5.1 Continuity

The central theme of analysis is that of a continuous function. Whenever possible we have defined our concepts in terms of sequences and, for the moment, we shall use sequences to define continuity. We saw in the first section of Chapter 3 that a function $f: \mathbb{R}^2 \to \mathbb{R}$ is continuous if whenever

$$(x_1, y_1), (x_2, y_2), (x_3, y_3), \ldots \to (x_0, y_0) \text{ in } \mathbb{R}^2$$

it follows that

$$f(x_1, y_1), f(x_2, y_2), f(x_3, y_3), \ldots \to f(x_0, y_0) \text{ in } \mathbb{R}.$$

That idea extends easily to a function $f: X \to X'$ where (X, d) and (X', d') are both metric spaces.

Definition Let (X, d) and (X', d') be metric spaces and let $f: X \to X'$. Then f is *continuous* if whenever

$$x_1, x_2, x_3, \ldots \to x_0 \text{ in } (X, d)$$

it follows that

$$f(x_1), f(x_2), f(x_3), \ldots \to f(x_0) \text{ in } (X', d')$$

Exercise 91 Let (X, d) and (X', d') be metric spaces and let $f: X \to X'$ have the property that for some fixed number L
$$d'(f(x), f(y)) \leqslant L d(x, y)$$
for all $x, y \in X$. Show that f is continuous.

Exercise 92 Let $f: C(0, 1) \to \mathbb{R}$ be given by $f(x) = x(0)$; i.e. for each member x of $C(0, 1)$, $f(x)$ is the value of that function at 0. Show that f is continuous (with respect to the usual metrics on $C(0, 1)$ and $\mathbb{R}$).

Exercise 93 Let (X, d) be a metric space and let $x_0 \in X$. Define $f: X \to \mathbb{R}$ by $f(x) = d(x_0, x)$. Show that, for each $x, y \in \mathbb{R}$,
$$|f(x) - f(y)| \leqslant d(x, y)$$
and deduce from Exercise 91 that f is continuous.

88 What makes analysis work?

We saw in Theorem 3.1 that if $f\colon \mathbb{R}^2 \to \mathbb{R}$ is continuous then $f^{-1}([a, b])$ is closed. This is a very special case of the following result.

Theorem 5.1 Let (X, d) and (X', d') be metric spaces and let $f\colon X \to X'$ be continuous. Then if A is closed in (X', d') it follows that $f^{-1}(A)$ is closed in (X, d).

Proof We must consider a convergent sequence in $f^{-1}(A)$ and show that its limit is also in $f^{-1}(A)$. So let

$$\underbrace{x_1, x_2, x_3, \ldots}_{\in f^{-1}(A)} \to x_0.$$

Then, by the continuity of f,

$$\underbrace{f(x_1), f(x_2), f(x_3), \ldots}_{\in A} \to f(x_0).$$

But A is closed and hence $f(x_0) \in A$ and $x_0 \in f^{-1}(A)$ as required. □

Exercise 94 Use Theorem 5.1 and Exercise 93 to show that if (X, d) is a metric space with $x_0 \in X$ and if r is a real number, then the sets

$$\{x \in X\colon d(x, x_0) \geq r\} \quad \text{and} \quad \{x \in X\colon d(x, x_0) \leq r\}$$

are closed in (X, d).

It turns out that the converse of Theorem 5.1 is also true, namely that if $f\colon X \to X'$ is such that $f^{-1}(A)$ is closed in (X, d) whenever A is closed in (X', d'), then f is continuous. It is therefore possible to characterise continuity in terms of closed sets.

Theorem 5.2 Let (X, d) and (X', d') be metric spaces. Then $f\colon X \to X'$ is continuous if and only if $f^{-1}(A)$ is closed in (X, d) whenever A is closed in (X', d').

Proof The fact that continuous functions have the stated property was established in Theorem 5.1. To prove the converse, assume that $f^{-1}(A)$ is closed in (X, d) whenever A is closed in (X', d'). To show that f is continuous we take a convergent sequence

$$x_1, x_2, x_3, \ldots \to x_0 \quad \text{in } (X, d)$$

and assume that

$$f(x_1), f(x_2), f(x_3), \ldots \not\to f(x_0) \quad \text{in } (X', d').$$

Then some subsequence of $f(x_1), f(x_2), f(x_3), \ldots$ fails to get close to

Continuity

$f(x_0)$; i.e. there exists an $r>0$ and a subsequence $f(x_{k_1}), f(x_{k_2}), f(x_{k_3})$, ... all of whose terms are distance r or more from $f(x_0)$. Hence $f(x_{k_1}), f(x_{k_2}), f(x_{k_3}), \ldots \in \{x \in X' : d'(f(x_0), x) \geq r\} = A$, say, which by Exercise 94 is a closed set in (X', d'). Hence, by the assumed condition on f, $f^{-1}(A)$ is closed in (X, d). But

$$\underbrace{x_{k_1}, x_{k_2}, x_{k_3}, \ldots}_{\in f^{-1}(A), \text{ closed}} \to x_0$$

and so $x_0 \in f^{-1}(A)$ and $f(x_0) \in A$. That means that $d(f(x_0), f(x_0)) \geq r > 0$, which is a contradiction. Thus $f(x_1), f(x_2), f(x_3), \ldots \to f(x_0)$ and f is continuous as claimed. □

Our definition and characterisation of continuity fits in with our restricted use of the concept in earlier chapters. But in a traditional approach to metric spaces the starting point is the ε–δ definition of continuity of a real function:

'$f: \mathbb{R} \to \mathbb{R}$ is continuous if given $x \in \mathbb{R}$ and $\varepsilon > 0$ there exists $\delta > 0$ such that

$|f(x) - f(y)| < \varepsilon$ whenever $|x - y| < \delta$;

i.e.

$f(y) \in]f(x) - \varepsilon, f(x) + \varepsilon[= I$ whenever $y \in]x - \delta, x + \delta[= J$.'

So traditionally a function is continuous if whenever I is an open interval containing $f(x)$ there exists an open interval J containing x with $f(J) \subseteq I$. In order to generalise this approach to functions from one metric space to another we need first to generalise the idea of an open interval. We could actually manage without this concept even in this last chapter, but in order that the reader can understand other texts we must include this generalisation here.

Definition Given a metric space (X, d), a member $x_0 \in X$ and a number $r > 0$, the *open ball* $B_X(x_0, r)$ is the set

$\{x \in X : d(x_0, x) < r\}$.

So, for example, the open ball $B_\mathbb{R}(x_0, r)$ (assuming the usual metric on $\mathbb{R}$) is the set

$\{x \in \mathbb{R} : |x_0 - x| < r\} =]x_0 - r, x_0 + r[$.

Exercise 95 Illustrate the open ball centre $(0, 0)$ and radius r in $\mathbb{R}^2$ in the following cases:

(a) with the usual metric,
(b) with the 'max metric' defined in Exercise 13, and
(c) with the 'lift metric' defined in Example **6** of Chapter 2.

The generalisation of our ε–δ definition of continuity to functions $f: X \to X'$ would now read

'$f: X \to X'$ is continuous if given $x \in X$ and $\varepsilon > 0$ there exists $\delta > 0$ such that

$f(y) \in B_{X'}(f(x), \varepsilon)$ whenever $y \in B_X(x, \delta)$.'

But it is actually possible to use the idea of an open ball to characterise continuity without the drawback of the technical complications of εs and δs.

Definition In a metric space (X, d) the set A is *open* if given $a \in A$ there exists $r > 0$ with $B_X(a, r) \subseteq A$.

In other words, A is open if each $a \in A$ is the centre of some open ball contained entirely in A. For example, in $\mathbb{R}$ the set $A =]0, \infty[$ is open because if $a \in]0, \infty[$, then the open ball $B_\mathbb{R}(a, a)$ ($=]0, 2a[$) is contained in A. An open set in $\mathbb{R}^2$ (usual metric) is illustrated in Figure 5.1.

Fig. 5.1

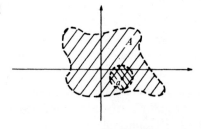

Exercise 96 Show that the set $\{(x, y): x > 0 \text{ and } y > 0\}$ is open in $\mathbb{R}^2$ (usual metric).

Exercise 97 Let A be the open ball $B_X(x, r)$ in the metric space (X, d). Use the triangle inequality to show that if $a \in A$ then $r - d(x, a) > 0$ and $B_X(a, r - d(x, a)) \subseteq A$. Deduce that open balls are open.

We can now characterise continuity in terms of open sets. This has the advantage of avoiding εs and δs. (Indeed, it avoids any mention of

Continuity

distance and it prepares the way for the generalisation – not covered in this text – to 'topology' where the basic concept is not distance but the idea of an open set.) The characterisation says that $f: X \to X'$ is continuous if and only if $f^{-1}(A)$ is open in (X, d) whenever A is open in (X', d'), and we (or you!) will prove this in Exercise 100.

Some readers will surely have already asked themselves how we have managed without the concept of an open set if, as we claim, it is a central idea in analysis. Other readers will have noticed the strong connection between the above statement concerning continuity in terms of open sets and the statement of Theorem 5.2 concerning continuity in terms of closed sets. The 'missing link' will be established in Theorem 5.3, but before stating that result we warn the reader not to think of open as 'not closed'.

> **Exercise 98** Let $A = [0, 1[$ in $\mathbb{R}$ (usual metric). Find a convergent sequence in A whose limit is not in A, and show also that $B_\mathbb{R}(0, r) \not\subseteq A$ for any $r > 0$. Deduce that A is neither open nor closed.
>
> **Exercise 99** Let (X, d) be a metric space. Show that X and $\emptyset$ are both open and closed in (X, d).

Theorem 5.3 Let (X, d) be a metric space. Then $A \subseteq X$ is open if and only if its complement $X \backslash A$ is closed (Figure 5.2).

Fig. 5.2

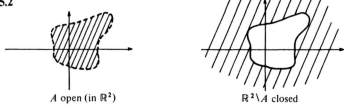

A open (in $\mathbb{R}^2$) $\mathbb{R}^2 \backslash A$ closed

Proof Assume first that A is open and let $x_1, x_2, x_3, \ldots$ be a sequence of points of $X \backslash A$ with $x_1, x_2, x_3, \ldots \to x_0$. To show that $X \backslash A$ is closed we must show that $x_0 \in X \backslash A$. Let us assume that $x_0 \notin X \backslash A$ and derive a contradiction. If $x_0 \notin X \backslash A$, then x_0 is in the open set A and so $B_X(x_0, r) \subseteq A$ for some $r > 0$. Hence

$$\underbrace{x_1, x_2, x_3, \ldots}_{\in X \setminus A} \to x_0$$
$$\subseteq X \setminus B_X(x_0, r)$$
$$= \{x \in X : d(x_0, x) \geq r\} \quad \text{(closed)}$$

and it follows that $x_0 \in \{x \in X : d(x_0, x) \geq r\}$; i.e. $d(x_0, x_0) \geq r > 0$, which is the required contradiction. Hence $X \setminus A$ is closed.

Conversely assume that A is not open. Then there exists an $a \in A$ with
$$B_X(a, 1) \not\subseteq A, \quad B_X(a, \tfrac{1}{2}) \not\subseteq A, \quad B_X(a, \tfrac{1}{3}) \not\subseteq A, \ldots;$$
i.e. there exist an $x_1 \in X \setminus A$ with $d(a, x_1) < 1$, an $x_2 \in X \setminus A$ with $d(a, x_2) < \tfrac{1}{2}$, an $x_3 \in X \setminus A$ with $d(a, x_3) < \tfrac{1}{3}, \ldots$. Clearly we have a sequence $x_1, x_2, x_3, \ldots$ in $X \setminus A$ convergent to $a \notin X \setminus A$. Hence $X \setminus A$ is not closed, and the theorem follows. □

Whenever we have needed a result concerning open sets in the earlier chapters we have therefore been able to quote the results in terms of closed sets.

Exercise 100 Use Theorems 5.2 and 5.3 to show that $f: X \to X'$ is continuous if and only if $f^{-1}(A)$ is open in (X, d) whenever A is open in (X', d').

Exercise 101 Use Theorems 3.2, 3.3 and 5.3 to show that
(a) any union of open sets in a metric space is itself open;
(b) if $A_1, A_2, \ldots, A_k$ are open sets in a metric space then $A_1 \cap A_2 \cap \ldots \cap A_k$ is open.

Give an example to show that an intersection of open sets need not be open.

We have shown the link between our approach (via sequences) and the traditional approach (via open sets). We are now able to look at the classical theorems concerning continuous functions and to generalise them to arbitrary metric spaces. At the same time we hope that this abstraction enables us to see in the underlying structure why the theorems work.

5.2 Attained bounds

Here is a result from introductory courses on real analysis:

Attained bounds

the proof is chosen as typical of the way in which the results are proved at that level.

Theorem 5.4 Let $f: [a, b] \to \mathbb{R}$ be continuous. Then f is bounded and attains its bounds; i.e. there exist numbers m, M with $m \leqslant f(x) \leqslant M$ for all $x \in [a,b]$ and numbers $\xi, \eta \in [a,b]$ with $m = f(\xi)$ and $M = f(\eta)$ (Figure 5.3).

Fig. 5.3

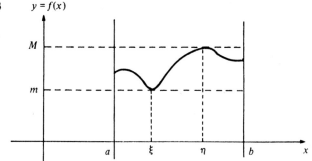

Proof We will show that the set $f([a, b]) = \{f(x): x \in [a, b]\}$ is bounded above and that there exists $\eta \in [a, b]$ with

$$f(\eta) = M = \sup f([a, b]);$$

the result for lower bounds follows similarly.

Assume then that $f([a, b])$ is not bounded above. Then there exists $x_1 \in [a, b]$ with $f(x_1) \geqslant 1$, $x_2 \in [a, b]$ with $f(x_2) \geqslant 2$, $x_3 \in [a, b]$ with $f(x_3) \geqslant 3, \ldots$.

Now $x_1, x_2, x_3, \ldots$ is a sequence in $[a, b]$ and so it has a convergent subsequence

$$x_{k_1}, x_{k_2}, x_{k_3}, \ldots \to x_0$$

in $[a, b]$. Therefore

$$f(x_{k_1}), f(x_{k_2}), f(x_{k_3}), \ldots \to f(x_0)$$
$$\geqslant k_1 \quad \geqslant k_2 \quad \geqslant k_3$$

⎫ by the compactness of $[a, b]$

⎬ by the continuity of f

whereas it is clear that the sequence $f(x_{k_1}), f(x_{k_2}), f(x_{k_3}), \ldots$ is tending to infinity. This contradiction shows that $f([a, b])$ is bounded above.

Let $M = \sup f([a, b])$. Then there is a sequence in $f([a, b])$ tending to M,

$f(y_1), f(y_2), f(y_3), \ldots \to M$.

Now $y_1, y_2, y_3, \ldots$ is a sequence in $[a, b]$ and so it has a convergent subsequence $\quad$ *by the compactness of $[a, b]$*

$$y_{l_1}, y_{l_2}, y_{l_3}, \ldots \to \eta$$

in $[a, b]$. Therefore

$$f(y_{l_1}), f(y_{l_2}), f(y_{l_3}), \ldots \to f(\eta).$$ *by the continuity of f*

But $f(y_1), f(y_2), f(y_3), \ldots \to M$ and so, by the uniqueness of limits, $M = f(\eta)$ as required. □

It seems that the only properties of $[a, b]$ and f used in that proof were the compactness of $[a, b]$ and the continuity of f. So the obvious generalisation is as stated in the next exercise.

> ***Exercise 102*** Let (X, d) be a compact metric space and let $f: X \to \mathbb{R}$ be continuous. Imitate the proof of Theorem 5.4 to show that there exist $\xi, \eta \in X$ with $f(\xi) \leqslant f(x) \leqslant f(\eta)$ for all $x \in X$.

That generalisation was no harder to prove than the more restricted version. The pleasant bonus to finish this section is that by generalising the result further the proof becomes easier.

Theorem 5.5 Let (X, d) and (X', d') be metric spaces with (X, d) compact. Let $f: X \to X'$ be continuous. Then $f(X) = \{f(x): x \in X\}$ is compact (and hence closed and bounded) in (X', d').

Proof We must take any sequence in $f(x)$ and find a convergent subsequence with its limit in $f(X)$. So let $f(x_1), f(x_2), f(x_3), \ldots$ be a sequence in $f(X)$. Then the sequence $x_1, x_2, x_3, \ldots$ is in the compact set X and so it has a convergent subsequence

$$x_{k_1}, x_{k_2}, x_{k_3}, \ldots \to x_0$$

in X. Hence, by the continuity of f,

$$f(x_{k_1}), f(x_{k_2}), f(x_{k_3}), \ldots \to f(x_0)$$

in $f(X)$. We have found the required convergent subsequence and thus established that $f(X)$ is compact. □

5.3   Uniform continuity

Returning again to the ε–δ definition of continuity, a function $f: [a, b] \to \mathbb{R}$ is continuous if for each $x \in [a, b]$ and each

Uniform continuity

$\varepsilon > 0$ there exists $\delta > 0$ (which may depend upon the choice of x and ε) such that (for $y \in [a, b]$)

$$f(y) \in \,]f(x) - \varepsilon, f(x) + \varepsilon[\quad \text{whenever } y \in \,]x - \delta, x + \delta[$$

or

$$|f(x) - f(y)| < \varepsilon \quad \text{whenever } |x - y| < \delta.$$

Sometimes it is possible to choose a δ which does not depend on x; i.e. given $\varepsilon > 0$ there exists $\delta > 0$ such that (for all $x, y \in [a, b]$)

$$|f(x) - f(y)| < \varepsilon \quad \text{whenever } |x - y| < \delta.$$

Such a function is said to be *uniformly continuous*.

A theorem from elementary analysis courses is that if $f: [a, b] \to \mathbb{R}$ is continuous then it is uniformly continuous. Analysts who approach metric spaces with a view to generalising that result soon see that again it is the compactness of $[a, b]$ which makes it work. But they start considering sets like $]x - \delta, x + \delta[$ covering $[a, b]$ and that leads them to consider compactness in the following way: 'a set A is compact in a metric space if and only if whenever A is contained in a union of open sets it follows that A is contained in the union of a finite number of them.' This characterisation (which turns out to be equivalent to our definition) has the advantage that it carries over to the further generalisation of 'topology' where all concepts have to be defined in terms of open sets. However, our form of compactness is still sufficient to enable us to understand why the result on uniform continuity works.

Theorem 5.6 Let $f: [a, b] \to \mathbb{R}$ be continuous. Then it is uniformly continuous.

Proof Assume that f is continuous but not uniformly continuous: we shall deduce a contradiction. Since f is not uniformly continuous there must exist some $\varepsilon > 0$ (which is fixed from this point onwards) so that *no* $\delta > 0$ has the property that (for $x, y \in [a, b]$)

$$|f(x) - f(y)| < \varepsilon \quad \text{whenever } |x - y| < \delta.$$

So

$\delta = 1$ fails; i.e. there exist $x_1, y_1 \in [a, b]$ with
$$|x_1 - y_1| < 1 \quad \text{but } |f(x_1) - f(y_1)| \geq \varepsilon.$$

$\delta = \tfrac{1}{2}$ fails; i.e. there exist $x_2, y_2 \in [a, b]$ with
$$|x_2 - y_2| < \tfrac{1}{2} \quad \text{but } |f(x_2) - f(y_2)| \geq \varepsilon.$$

96   What makes analysis work?

$\delta = \frac{1}{3}$ fails; i.e. there exist $x_3, y_3 \in [a, b]$ with $|x_3 - y_3| < \frac{1}{3}$ but $|f(x_3) - f(y_3)| \geq \varepsilon$.

$\vdots$

The sequence $x_1, x_2, x_3, \ldots$ in $[a, b]$ must have a convergent subsequence

$$x_{k_1}, x_{k_2}, x_{k_3}, \ldots \to x_0$$

in $[a, b]$. Since

$$|x_{k_n} - y_{k_n}| \to 0$$

it follows that

$$y_{k_1}, y_{k_2}, y_{k_3}, \ldots \to x_0$$

also. Hence

$$f(x_{k_1}), f(x_{k_2}), f(x_{k_3}), \ldots \to f(x_0)$$

and

$$f(y_{k_1}), f(y_{k_2}), f(y_{k_3}), \ldots \to f(x_0).$$

⎫ by the compactness of $[a, b]$
⎫ by the triangle inequality
⎫ by the continuity of f

Therefore $|f(x_{k_n}) - f(y_{k_n})| \to 0$, which contradicts the fact that $|f(x_n) - f(y_n)| \geq \varepsilon > 0$ for all n. Hence the continuity of f does imply its uniform continuity. □

Definition Let (X, d) and (X', d') be metric spaces. Then $f: X \to X'$ is *uniformly continuous* if given $\varepsilon > 0$ there exists $\delta > 0$ such that, for all $x, y \in X$,

$$d'(f(x), f(y)) < \varepsilon \quad \text{whenever } d(x, y) < \delta.$$

Exercise 103 Let (X, d) and (X', d') be metric spaces with (X, d) compact. Imitate the proof of Theorem 5.6 to show that if $f: X \to X'$ is continuous then it is uniformly continuous.

5.4   Inverse functions

Here is another standard result and proof from courses in real analysis.

Theorem 5.7 Let $A \subseteq \mathbb{R}$ and let $f: [a, b] \to \mathbb{R}$ be continuous and bijective (i.e. for each $y \in A$ there is precisely one $x \in [a, b]$ with $f(x) = y$). Then f has a well-defined inverse $f^{-1}: A \to [a, b]$ and it, too, is continuous (Figure 5.4).

Inverse functions

Fig. 5.4

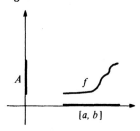
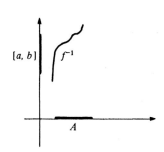

Proof The inverse function is certainly well defined because for each $y \in A$ there is just one $x \in [a, b]$ with $f(x) = y$: so $f^{-1}(y)$ is precisely that x.

To show that f^{-1} is continuous we let $a_1, a_2, a_3, \ldots \to a_0$ in A: we need to deduce that $f^{-1}(a_1), f^{-1}(a_2), f^{-1}(a_3), \ldots \to f^{-1}(a_0)$. Assume that this fails. Then some subsequence must fail to get close to $f^{-1}(a_0)$; i.e. there exists $r > 0$ and a subsequence $f^{-1}(a_{k_1}), f^{-1}(a_{k_2}), f^{-1}(a_{k_3}), \ldots$ where members are all distance at least r from $f^{-1}(a_0)$. The sequence

$$f^{-1}(a_{k_1}), f^{-1}(a_{k_2}), f^{-1}(a_{k_3}), \ldots$$

is in $[a, b]$ and so it has a convergent subsequence $\quad\Big\}\quad$ by the compactness of $[a, b]$

$$f^{-1}(a_{l_1}), f^{-1}(a_{l_2}), f^{-1}(a_{l_3}), \ldots \to \alpha$$

in $[a, b]$. Since the terms of this sequence have

$$|f^{-1}(a_{l_n}) - f^{-1}(a_0)| \geq r$$

it follows that $\quad\Big\}\quad$ since $\{x \in \mathbb{R} : |x - f^{-1}(a_0)| \geq r\}$ is closed

$$|\alpha - f^{-1}(a_0)| \geq r > 0$$

and that $\alpha \neq f^{-1}(a_0)$.

Now since

$$f^{-1}(a_{l_1}), f^{-1}(a_{l_2}), f^{-1}(a_{l_3}), \ldots \to \alpha$$

it follows that $\quad\Big\}\quad$ by the continuity of f

$$f(f^{-1}(a_{l_1})), f(f^{-1}(a_{l_2})), f(f^{-1}(a_{l_3})), \ldots \to f(\alpha)$$

i.e.

$$a_{l_1}, a_{l_2}, a_{l_3}, \ldots \to f(\alpha).$$

98 What makes analysis work?

But $a_1, a_2, a_3, \ldots \to a_0$ and so by the uniqueness of limits $f(\alpha) = a_0$ and $\alpha = f^{-1}(a_0)$. This contradicts the earlier statement and shows that f^{-1} is indeed continuous, as required. □

Exercise 104 Let (X, d) and (X', d') be metric spaces with (X, d) compact. Imitate the proof of Theorem 5.7 to show that if $f: X \to X'$ is continuous and bijective, then the inverse function $f^{-1}: X' \to X$ is also continuous.

The result quoted in Exercise 104 can, in fact, be proved in a rather more sophisticated fashion.

Theorem 5.8 Let (X, d) and (X', d') be metric spaces with (X, d) compact. Then if $f: X \to X'$ is continuous and bijective its inverse $f^{-1}: X' \to X$ is also continuous.

Proof To show that $f^{-1}: X' \to X$ is continuous we use the characterisation of continuity from Theorem 5.2 and show that if B is closed in (X, d) then $(f^{-1})^{-1}(B)$ $(= f(B))$ is closed in (X', d'). But

B closed in the compact (X, d)

$\Rightarrow B$ compact in (X, d) (Exercise 43)

$\Rightarrow f(B)$ compact in (X', d') (Theorem 5.5)

$\Rightarrow f(B)$ closed in (X', d') (Theorem 3.11)

$\Rightarrow (f^{-1})^{-1}(B)$ closed in (X', d').

Hence f^{-1} is continuous, as claimed. □

5.5 Intermediate values

In Chapter 3 we considered the 'three Cs', closed, complete and compact. There is a fourth equally important property of sets in a metric space, that of 'connectedness'. The reason that we have left this concept until now is that it played no role in our iterative approach.

Informally a set is 'connected' if it is in 'one piece'. So, for example, we would expect the set F (in $\mathbb{R}^2$) illustrated in Figure 5.5 to be connected, but the sets D and E to be disconnected.

To show that D is disconnected we write it as the union of two 'separate' pieces.

$$D = \{(x, y): x^2 + y^2 \leq 1\} \cup \{(x, y): x^2 + y^2 \geq 2\},$$

and similarly we write E as the union of two separate pieces

$$E = \{(x, y): x > 0 \text{ and } y \geq 0\} \cup \{(x, y): x < 0 \text{ and } y \leq 0\}.$$

Intermediate values

Fig. 5.5

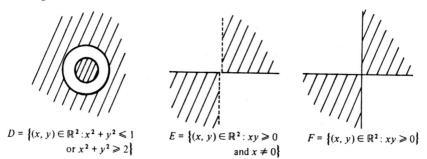

$D = \{(x, y) \in \mathbb{R}^2 : x^2 + y^2 \leq 1$
$\text{or } x^2 + y^2 \geq 2\}$

$E = \{(x, y) \in \mathbb{R}^2 : xy \geq 0$
$\text{and } x \neq 0\}$

$F = \{(x, y) \in \mathbb{R}^2 : xy \geq 0\}$

But one could argue that

$F = \{(x, y): x \geq 0 \text{ and } y \geq 0\} \cup \{(x, y): x \leq 0, y \leq 0 \text{ and } (x, y) \neq (0, 0)\}$

splits F into two separate pieces. What is it that distinguishes the split of D and E above from the split of F? In the cases of D and E the two parts can be surrounded by disjoint open sets (see Figures 5.6 and 5.7):

$$D = \{(x, y): x^2 + y^2 \leq 1\} \cup \{(x, y): x^2 + y^2 \geq 2\},$$
$$\cap \qquad\qquad\qquad \cap$$
$$\{(x, y): x^2 + y^2 < \tfrac{5}{4}\} \quad \{(x, y): x^2 + y^2 > \tfrac{7}{4}\}$$

open and disjoint

$$E = \{(x, y): x > 0 \text{ and } y \geq 0\} \cup \{(x, y): x < 0 \text{ and } y \leq 0\}.$$
$$\cap \qquad\qquad\qquad \cap$$
$$\{(x, y): x + y > 0\} \qquad \{(x, y): x + y < 0\}$$

open and disjoint

Fig. 5.6

Fig. 5.7

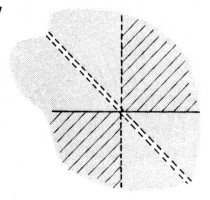

Fig. 5.8

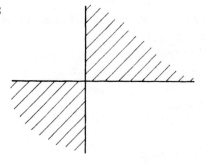

However, we cannot do this in the case of our split of F (see Figure 5.8):

$$F = \underbrace{\{(x,y): x \geq 0 \text{ and } y \geq 0\}}_{B} \cup \{(x,y): x \leq 0, y \leq 0 \text{ and } (x,y) \neq (0,0)\}$$

If B is open then, as $(0,0) \in B$, it follows that a ball centre $(0,0)$ lies entirely in B. So B must clearly overlap the other part of F.

Definition In a metric space (X, d) a set A is *disconnected* if $A = A_1 \cup A_2$ where A_1, A_2 are non-empty and $A_1 \subseteq B$, $A_2 \subseteq C$ for some disjoint open sets B and C. Otherwise A is *connected*.

As you might expect, the connected sets in $\mathbb{R}$ are precisely the intervals ($[0,1],]2,4],]5,9[,]1,\infty[, \mathbb{R}$, etc.) which are characterised by the fact that if they contain two points then they contain all points between them.

Intermediate values

Theorem 5.9 In $\mathbb{R}$ (with the usual metric) the connected sets are precisely the intervals.

Proof Assume first that $A \subseteq \mathbb{R}$ is disconnected. Then $A = A_1 \cup A_2$ with $b \in A_1$, $c \in A_2$ and $b < c$, say, and with $A_1 \subseteq B$, $A_2 \subseteq C$ for some disjoint open sets B and C. The set $[b, c] \cap B$ is non-empty and bounded above and so it has a supremum ξ. Now $\xi \notin C$, for if it were in that open set then there would exist $r > 0$ with $]\xi - r, \xi + r[\subseteq C$. That would make $\xi - r$ an upper bound for $[b,c] \cap B$ and would contradict the choice of ξ as the least upper bound. Similarly $\xi \notin B$, for if it were in that open set there would exist $s > 0$ with $]\xi - s, \xi + s[\subseteq B$ which (together with the fact that $\xi < c$) would contradict the fact that ξ is an upper bound of $[b,c] \cap B$. Hence $\xi \notin B \cup C$ and as $A \subseteq B \cup C$ it follows that $\xi \notin A$. Hence $b, c \in A$ but $\xi \in [b,c]$ and $\xi \notin A$: it follows that A is not an interval.

Conversely if $A \subseteq \mathbb{R}$ is not an interval then there exist $b, c \in A$ and an $a \in]b, c[$ with $a \notin A$. Then

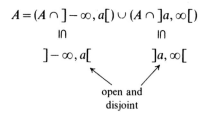

shows that A is disconnected.

Hence A is disconnected if and only if it is not an interval, and the theorem follows. $\square$

We are now able to recall the most classic result of elementary analysis: we only give an outline of the sort of proof encountered in elementary courses.

Theorem 5.10 (The intermediate value theorem.) Let $f: [a, b] \to \mathbb{R}$ be continuous and let c be a number between $f(a)$ and $f(b)$. Then $c = f(\xi)$ for some $\xi \in [a,b]$; i.e. f takes all 'intermediate values'.

What makes analysis work?

Outline proof Assume that $f(a) < c < f(b)$ and let $B = \{x : f(x) < c\}$ (see Figure 5.9).

Fig. 5.9

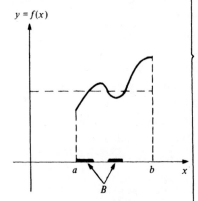

Then B has a supremum ξ and (by various continuity arguments) it can be shown that $\xi \notin B$ and $\xi \notin C = \{x : f(x) > c\}$. Hence $f(\xi) = c$ as required. $\square$

Basically this is choosing a number in $[a, b]$ but not in the disjoint open sets $B = f^{-1}(]-\infty, c[)$ and $C = f^{-1}(]c, \infty[)$; i.e. it is using the connectedness of $[a, b]$.

Exercise 105 Let (X, d) be a connected metric space (i.e. X is itself connected) and let $f: X \to \mathbb{R}$ be continuous. Assume that $a, b \in X$ and $c \in \mathbb{R}$ are such that $f(a) < c < f(b)$. Show that there exists $\xi \in X$ with
$$\xi \notin f^{-1}(]-\infty, c[) \cup f^{-1}(]c, \infty[).$$
Deduce that $f(\xi) = c$.

Our final result generalises Theorem 5.10 and Exercise 105 and, at the same time, its proof becomes most straightforward and natural.

Theorem 5.11 Let (X, d) and (X', d') be metric spaces with (X, d) connected. Then if $f: X \to X'$ is continuous it follows that $f(X) = \{f(x) : x \in X\}$ is connected in (X', d').

Proof We shall show that if $f(X)$ is disconnected then X is disconnected. Indeed, we shall show that if the open sets B, C

'disconnect' $f(X)$ then the open sets $f^{-1}(B)$, $f^{-1}(C)$ will 'disconnect' X.

Assume then that
$$f(X) = A_1 \cup A_2 \quad \text{(non-empty)}$$
$$\cap \quad \cap$$
$$B \quad C.$$

open and disjoint in (X', d')

Then, by the continuity of f, the sets $f^{-1}(B)$ and $f^{-1}(C)$ are open in (X, d). Also they are disjoint since
$$f^{-1}(B) \cap f^{-1}(C) = f^{-1}(B \cap C) = f^{-1}(\emptyset) = \emptyset.$$
Finally
$$X = f^{-1}(f(X)) = f^{-1}(A_1 \cup A_2) = f^{-1}(A_1) \cup f^{-1}(A_2)$$
$$\cap \quad \cap$$
$$f^{-1}(B) \quad f^{-1}(C).$$

open and disjoint

So if X is connected it follows that $f(X)$ is connected. □

5.6 Some final remarks

I hope that the reader who has persevered this far will have found the task rewarding. Not everyone appreciates mathematical abstraction and for that reason our whole approach was motivated by iterative techniques. However, having set up all the machinery for that approach, it was then an easy step for us to use that machinery to look with fresh insight at analysis.

Those readers who wish to pursue the subject further (perhaps towards functional analysis or to topology) can read any of the excellent books on the subject. In particular the following three have given me a great deal of pleasure and I can recommend them to you. The first covers a little more material than our approach, but the other two go much further.

Metric Spaces, by E. T. Copson (Cambridge University Press, 1968).

Introduction to Topology and Modern Analysis, by G. F. Simmons (McGraw Hill, 1963).

Introduction to Metric and Topological Spaces, by W. A. Sutherland (Oxford University Press, 1975).

Finally, I hope that you have enjoyed this natural approach to a piece of abstract mathematics, and I wish you well in your future reading.

INDEX

attained bound 92–4
Banach's fixed point principle 58–60
bibliography 103, 104
bounded sequence 39
bounded set 21, 48
$C(a,b)$ 20
Cauchy criterion 39
Cauchy sequence 39
closed interval 19
closed set 30–5
compact set 45–51
compact space 69
completeness of $C(a,b)$ 42
completeness of $\mathbb{R}$ 39
completeness of $\mathbb{R}^2$ 41
complete set 38–45
complete space 40, 45
connected 100, 101
continuous 5, 31, 87
convergent 3, 23
coordinatewise convergence 24, 26–8
contraction 57
contraction mapping principle 58–60

differential equation 68, 71–83
disconnected 100
discrete space 16
distance 12, 13

Euclidean distance 14

fixed point 2, 56

half-plane 33

implicit function 84
implicit function theorem 84
integral equation 10, 68
intermediate values 98
intermediate value theorem 101

inverse function 96–8
irrational 30
iterate 64
iteration 2, 31

least upper bound 21
lift metric 17, 18
limit 3, 23, 28
linear algebra 61, 62

matrices 61, 62
max metric on $C(a,b)$ 19, 20
max metric on $\mathbb{R}^n$ 18
mean value theorem 63
metric 13
metric space 13
metric (usual) on $C(a,b)$ 19
metric (usual) on $\mathbb{R}$, $\mathbb{C}$ and $\mathbb{R}^2$ 14

Newton–Raphson 4

open ball 89
open interval 10
open set 35, 90

rational 29, 30

sequence 2, 22
self-counting list 7
simultaneous differential equations 80–3
simultaneous equations 57, 58, 60
subsequence 28
sup metric 21
supremum 21

topology 35, 91
triangle inequality 13

uniformly continuous 94–6
uniformly convergent 28
unique fixed point 56, 58

Printed in the United States
6936